THE
OLD TOWN CANOE
COMPANY

Our First Hundred Years

THE
OLD TOWN CANOE
COMPANY

Our First
Hundred Years

Susan T. Audette

with *David E. Baker*

Tilbury House, Publishers
Gardiner, Maine

Tilbury House, Publishers
132 Water Street
Gardiner, Maine 04345
800-582-1899

First printing: October, 1998

10 9 8 7 6 5 4 3 2 1

CIP Data:
Audette, Susan T., 1948–
The Old Town Canoe Company : our first hundred years / Susan T. Audette with David E. Baker.
p. cm.
Includes bibliographical references and index.
ISBN 0-88448-202-2 (hc : alk. paper). — ISBN 0-88448-203-0 (pb : alk. paper)
1. Old Town Canoe Company—History. 2. Boating industry—United States—History. 3. Canoes and canoeing—United States—Equipment and supplies—History. 4. Kayaks—United States—History. 5. Kayaking—United States—Equipment and supplies—History.
I. Baker, David E., 1941– . II. Title.
HD9993.B6340433 1998.
338.7'623829—dc21   98–41658
                                                      CIP

Designed on Crummet Mountain by Edith Allard, Somerville, Maine
Editorial and Production: Jennifer Elliott, Barbara Diamond, Mackenzie Dawson
Layout: Nina Medina, Basil Hill Graphics, Somerville, Maine
Image Scanning: Integrated Composition Systems, Benson Gray, Jennifer Elliott
Final Film: Integrated Composition Systems, Spokane, Washington
Printing and Binding: Worzalla Publishing, Stevens Point, Wisconsin

To my husband Vinny,
who bought me my first OLD TOWN canoe
and whose constant encouragement is
only outweighed by his patience.

Acknowledgments

The journey to recapture OLD TOWN's history has brought me many places and I have met many people along the way. Their contributions have been invaluable. I would like to extend my thanks to the following people: to Dave Baker, whose knowledge of canoe history and forestry has helped put things in proper perspective and who added a historical dimension that otherwise would be missing. Dave was a welcome source of ideas and energy throughout the project. To his wife Scottie, for her suggestions and support, as well as her hospitality during the many nights we spent laboring over the script. To the Gray family, who so willingly let me into their life story—especially Benson Gray, without whom this book would not exist. To Jim and Pinkie Cunningham for their patience with my never ending questions. To Linwood Wickett, whose quest for more information on his genealogy complemented my own search for answers. To Bob Pollock, whose work and love of Norumbega Park has added richness to this text. To Paula Hammett and Dan Butler for their help in filling in—providing time for me to work on the book. To Willie Hedricks of Valley Park, Missouri, a friend of Alfred Wickett, for his insight. To Sam Johnson and Peter Ottowitz, who took time from their busy schedules to meet with me. To the many workers and people of OLD TOWN, past and present, who have shared their labor of love, especially Lew Gilman, Geoff King, Joe LaVoie, Emedy Baillargeon, Durwood Pelky,

John Blass, Steve Krautkremer, Scott Phillips, Bart Hauthaway, John Whitney, Ethel Jenkins, James Day, Jim Stephenson, Steve Goslin, Frank Loring, Wendall Easler, Jerry Stelmok, Zip Kellogg, Ed Osgood, Pearl Veazie, Elizabeth Buck, Ralph Bouchard, Ruby Lutes, Genevieve Violette, William Dutch, Linwood Wickett, Sr., Bob Cardin, Roger Bills, Jeanne Wollsteadt and the staff at the Old Town Museum, and anyone else I may have forgotten to mention. To the many librarians who helped me, including Val Osgood and her staff at the Old Town Library, who constantly sent materials back and forth to Connecticut for my use, to the staff of the Special Collections Department at the University of Maine's Fogler Library, and the staff at the Daytona Racing Archives who were so helpful in obtaining photographs. To Cynthia Curtis at WoodenBoat, Jerry Cassell at The Wilderness Collection, Peter Vermilya at Mystic Seaport Museum, Llewellyn Howland III at Howland & Company, Hallie Bond at the Adirondack Museum, and Rollin Thurlow for their suggestions and assistance. To the many people who shared anecdotes with me for this book, their intimate stories telling what paddling is all about. To the staff at Tilbury House, Publishers for their patience, especially Jennifer Elliott, who always saw the importance of this project. Finally, to Paul and Helen Reagan, who provided a laugh whenever I needed one far from home.

—Sue Audette

C O N T E N T S

I began this adventure eight years ago after listening to Mike Hanna, a respected canoe builder, who was lecturing at a Wooden Canoe Heritage Association meeting at Lake Sebago in Maine. (WCHA is dedicated to the preservation and history of the wooden canoe.) He chastised the audience saying, "What a shame that we didn't capture the oral histories of all the great canoe makers that have come and gone." Because an OLD TOWN CANOE has been so much a part of my life and the lives of so many of us who have spent time in a canoe, I thought I should take Mike up on his challenge.

This story is written from a people perspective—you won't find every fact on each model that you might be looking for. Instead, I have spoken to many people connected with this great company and have learned about their lives and work. For me, they validated what I knew all the time—an OLD TOWN craft is a beautiful piece of work, and they were proud to be a part of it. Even though their work was a job, day in and day out they did it well and loved it. They worked hard to please the customer, even when they didn't necessarily agree with him. There wasn't always time for perfection and occasionally they made mistakes, but they loved their work.

What also became really clear to me is that this century-old business *was* and *is* not afraid to try something new. Although OLD TOWN CANOE COMPANY was tied to wood construction for many years, in modern times and with incredible effort, it has reached out into new waters. Most of its risk-taking has turned out to be extremely successful (and some is better left forgotten). The OLD TOWN canoes, kayaks, and boats were always made *by* people *for* people—and they still are.

Mike Hanna passed away three years following that assembly. I never got a chance to thank him.

—Susan T. Audette

PROLOGUE

It has been many years since the first canoes plied the waters. The place is different now. The town is one like so many throughout our country that has undergone change. Several storefronts are empty but their weathered boards and faded paint seem to hold a story of different times, more prosperous ones. In the background, you hear the rushing river that once held many canoeists, each with a purpose and goal. The river and the canoe are a natural combination; their union here produced an extraordinary story.

My Old Town Canoe

Up where the placid waters shine
Where men live close to God
I met you first old friend of mine
Out with my line and rod.

How often in the silent night
Just you and I would go
And try hard for a single bite
From the finny folks below.

And there beneath a friendly birch
We've fished close to the shore
And pulled out bass or wiggling perch
That weighed three pounds or more.

And if by chance I'd lose a perch
Or one would wiggle free
It seems that you would rock and lurch
Just making fun of me.

Your friendship's like the Sons of Maine
It's constant ever true
A better friend I'd seek in vain
Than you, Old Town Canoe...

THE

OLD TOWN CANOE
COMPANY

Our First Hundred Years

Old Town is a quiet, unassuming place these days, typical of many inland towns in Maine. Its two major employers are a paper mill that sits on the banks of the Penobscot River, and a canoe company tucked a city block away from the intersection of the main streets, where seldom-used railroad tracks crisscross the roads. It's hard to believe that Old Town was once the hub of a lumbering frenzy where huge quantities of raw logs were turned into lumber and fortunes were made. The river is no longer crowded with log booms and river drivers, but a hundred years ago it spawned another industry that grew steadily as the logging business declined. The OLD TOWN CANOE COMPANY is now the world's largest manufacturer of canoes and kayaks.

Our story of the OLD TOWN CANOE COMPANY must begin with the Native Americans who originated, perfected, and adopted the birchbark canoe as their watercraft, particularly the Woodland Indians called Abenaki, of Algonquian (var. sp.) stock, who occupied the North-Northeast at the time of European contact. Their name, which means "people of the Dawnland" or simply "Easterners," refers to their "nearness to the rising sun."[1]

The Abenaki were and are composed of two main branches, the larger of which is the eastern Abenaki, who lived primarily in Maine in distinct bands usually known by the names of the rivers near their settlements.[2] The Penobscot tribe has inhabited what is now Old Town since before recorded history.[3] The name "Old Town," originally the Indian village on Indian Island, indicated the first town, the earliest.[4]

1

A People, A River & a New Town

According to Fannie Hardy Eckstorm in her 1941 book *Indian Place-Names of the Penobscot Valley and the Maine Coast*, the name Penobscot means "at the descending rock."[5] The term originally identified "an area of about ten miles of the river between Bangor and Old Town,"[6] but eventually the entire river from its headwaters deep in the wilderness embracing Katahdin, Maine's highest peak, to the river's mouth where it enters Penobscot Bay at the Atlantic Ocean, would become known as the Penobscot, as would the Indian tribe that inhabited the area, a place that had provided for their needs for hundreds of years. They embraced it, protected it, and called it their own. It was their paradise.

The Penobscots were seasonally migratory people. In the summers, they traveled in small family bands to the coast to fish, returning to the interior to hunt in the winter. The waterways were their highways, and their birchbark canoes had evolved into superb, lightweight watercraft, built from the natural woodland materials around them: cedar and bark, lashed and sewn with spruce roots, the seams sealed with pitch.

The first white men to skirt the coast of Maine are thought to have been Basque, Portuguese, and Scandinavian fishermen who were searching for codfish as early as the late 1400s. Sebastian Cabot's journal mentions that the codfish were so closely packed together in these northern waters that his ships were unable to sail through them.[7] Stories of the unlimited bounty soon spread throughout western Europe. Companies were formed in Spain, England, Holland, and France to fish from Newfoundland and Nova Scotia southward, inevitably bringing more contact between the Indians and Europeans.

Columbus's well-known voyage of 1492 and the spectacular success of the Spaniards in conquering the Aztec Empire of Mexico inspired more and more Europeans to venture west in search of riches.[8] For many explorers, treasure hunters, and empire builders, rumors of a fabled, wealthy city called "Norumbega," where people adorned themselves with gold, silver, and pearls, fanned the flames of adventure even more. Well-known European trader-explorers kept the rumor alive. Giovanni da Verrazano, who landed at North Carolina in March of 1524 and then sailed as far north as Canada, named the entire area from the Gulf of St. Lawrence to the Hudson River "Norumbega," a name that was later applied to New England, then only to a part of Maine, to the paradise of the Penobscots.[9]

In 1579–1580 an English expedition to the Penobscot failed to find anything that could substantiate the rumors of a city called Norumbega. In 1604, Samuel de Champlain reached the Penobscot River where he met some Indians, and recorded in his journal, "Having made friends with them they guided us into their river.... I believe that this river is the one that several pilots and historians call Norumbega."[10] Champlain's companion, Lescarbot, wrote that the Spanish had been exaggerating about "a great and powerful city which they have named (neither I nor they know why) Norumbega.... If this fair town ever existed I would fain know who has destroyed it in the last eighty years; for there is nothing but scattered wigwams made of poles covered with bark or skin, and the name of both the settlement and the river is

Pemptegoet [Penobscot]."[11]

Despite the lack of golden treasures, the explorers were not blind to the resources within their grasp. In 1614 Captain John Smith was commissioned by a group of merchants to lead two ships to search for gold and silver or to hunt for whales. Smith found no precious metals but was instead struck by the desolation of Maine's coast and called it "a Countrie rather affright, then [sic] to delight one."[12] Yet, like those who followed, he was quick to see the Dawnland's potential:

> From Penobscot to Sagadahock this coast is all Mountainous and Isles of huge Rocks, but overgrown with all sorts of excellent good woods for building houses, boats, barks, or shippes; with an incredible abundance of most sorts of fish, much fowle, and sundry sorts of good fruits for man's use.[13]

These and other reports did not go unnoticed, and beginning in the first decade of the 1600s, efforts were made to establish settlements along the coast of Maine. The earliest were ill-fated, but more would follow.

In 1625 the Penobscot Indian population may have been as high as eight thousand, but their numbers decreased over time as they were invaded by uninvited settlers and their diseases, and by rival tribes. The Mohawk Indians attacked repeatedly between 1662–1669 and decimated them, but the Penobscots eventually established and maintained peace with the Mohawks. French Jesuits had visited them in the early 1600s, which led to the Penobscots' conversion and the establishment of a Catholic church on what is now called Indian Island in 1688.

Penobscot Indians at Indian Island gather for a photo, circa 1880s.
Old Town Public Library, Courtesy Fogler Library, Special Collections

Then in 1723 the unsuspecting Penobscots were besieged by a force of English Marines who drove the Indians away and burned the church and village wigwams, making hundreds homeless.[14] Despite the adversity, the Penobscots endured; they were anxious to live in peace and resume their life of reliance on the land. The Commonwealth of Massachusetts, which lay claim to what is now Maine, attempted to reserve some land for the Penobscots, but a series of treaties ultimately relieved the Indians of all their former lands except Indian Island, the other islands above it on the Penobscot River, and four townships

By 1833 the Penobscot population had dwindled to five hundred. The few remaining tribal members forged a treaty with the Commonwealth of Massachusetts, selling the four townships that remained in Indian hands for $50,000, which would be invested in the Commonwealth with interest being paid the tribe annually. When the Commonwealth divided, the newly formed State of Maine continued to recognize the treaty.[15]

try's population growing, especially on the East Coast, there was an insatiable demand for lumber. Speculators bought up the available land. "Cruisers" scrutinized the forest for the logging companies and loggers soon followed.[16] What awaited them were thousands of acres of virgin forests rich in hardwoods (oak, maple, ash, elm, basswood, poplar, and birch) and conifers (pine, cedar, hemlock, spruce, and fir). Here was their newest cache, "lumberman's gold."

The greatest obstacle was cutting the trees and getting the harvest out. Despite Maine's harsh weather and the isolated conditions in the interior, Mother Nature had made the logger's job easier by providing the Penobscot River.

River drivers—the men who took on the job of moving the logs from the forest downriver to the mills—never had it easy. They were a special breed: red-shirted, wool-clothed, strong, tough, callused, "stringy" men who were capable of remarkable feats. They were hardened by the cold weather and cold water, accustomed to the isolation, and feared nothing—least of all, death.

> It has always been the glory of the West Branch Drive that it had so many such men, every one of whom placed the welfare of those logs above his own life, could have handled the whole drive if there were need, and whose insubordination would never have gone so far as to endanger the least part of their trust.[17]

For all their toughness, they held a strong belief in God, embellished by their own stories and superstitions. One man who had seen this told the others:

These treaties opened the interior for exploitation. Until now, businessmen and loggers who had found wealth in timber had concentrated their efforts nearer the coastline. They had quickly stripped the southern areas and were now looking inland. With the coun-

I seen him stand there like he was on a barn floor, and I seen him lift up his fist an' shake it right stret in the face of old Katahdin, an' I hear him holler like his voice would rattle lead inside him, "To hell with God!"

An' then when I looked the Gray Rock was all empty an' in the water I seen only his two sets of fingers moving slow-like in the mist that sticks close to the black slick of the falls. I seen 'em open once, an' then they shut an' was gone...."

"That was a judgment for swearing," they answered solemnly, continuing their search for his body.

But the body was not to be found.

"And it ain't to be expected it ever will be. It ain't often that you do find 'em when they dies so by a judgment," said one of the wise ones who could remember much that had happened on the river.... it's always so with judgments; that's a part of it—they can't never be quiet till they are buried, and they don't never get buried, not that kind, when they die damning God that way.[18]

The approach of the "drive" to Old Town marked a high point in the year's lumbering activity. A cannon on Indian Island would be fired to announce its arrival, then sometimes as many as twenty bateaux filled with men could be seen racing down the river to unload at the landing. All respectable young ladies remained in their homes behind drawn curtains, either by order or through choice, as the logging crews returned to civilization after a stay of eight or nine months in the woods. One former Old Town woman, brave enough to watch the passing scene, described the sight:

Some of the men had not even shaved. Their clothing would have caused a scarecrow to throw up his hands in disgust. Red shirts were the most popular and conspicuous article of clothing visible until the wearer landed from the boats, then you couldn't tell what they were wearing for nether garments. Several layers of trousers, or in some cases, long woolen drawers topped the structure. The garment that boasted the fewest holes was worn on the outside. Some hats were without brims. Some were bareheaded because hat or cap with which the man had left home had blown away in some high wind and there was no means of replacement.[19]

The logs sent down the Penobscot by the river drivers bore the marks of many owners. The logs had to be collected, sorted by mark, and rafted together to be turned over to their owners. They were then scaled for estimated yield and sold before being sent on to the sawmills at Old Town and beyond.

At first the logs were corralled by boat crews working around the clock during the drive. Fires were lit along the shoreline at night to enable the men to see the logs as they approached. The process became more efficient in 1825, when the legislature granted a charter to a group of eighteen men who formed the Penobscot Boom Corporation. A "boom" was a line of connected, floating timbers stretching across the river to collect and keep the sawlogs together. Booms were anchored by stone-filled crib work piers (which can still be seen in many of Maine's big rivers). The first

attempt at collecting the sawlogs on the Penobscot was with just one boom, but it soon became evident that more were needed to handle the volume. Eventually a series of booms was erected, most often between an island and the mainland, along a ten-mile stretch of the river, and they became known collectively as the Penobscot Boom. The primary owners of the islands, the Penobscot Indians, capitalized on their landholdings and leased shore rights to the Boom.

The main booms stopped all the logs and had sorting gaps or chutes which allowed the manageable passage of logs. Crews working for the owners of various "marks" were stationed below these gaps to sort and raft their specific logs.

Each crew, interested in securing its own logs and rafting them, had a man standing on a log midstream. He kept his position by grasping a rope which was stretched across the water, the ends being fastened to trees, one on the shore and the other on the island. Standing with one foot on a log and holding the check line, the man could read the marks on the log, or, if the mark was not up, he could roll the log with his free foot until he could see the mark. He allowed to pass all logs without the mark he was interested in, but when his mark came, he gave the log a vigorous kick toward shore where the rafting crew secured it. In the days when two hundred men were working at the boom, these rafting crews filled the shore for nearly a mile.[20]

The original charter changed hands in 1827, was enlarged in 1832, and came under the sole ownership of General Samuel Veazie by 1834. General Veazie operated the boom until 1847, when he sold it to David Pingree, and it was still owned by the Pingree Estate in 1929. From 1854 until at least 1944, through an amendment to the charter by the legislature, the boom was leased to the Penobscot Lumbering Association, the lumbermen who used it.

In 1842 74,331,000 board feet of timber went through the boom; in 1855 that had increased to almost 182,000,000 board feet, with a 20-percent increase by 1882. Considering the logging methods of the day—crosscut saws, horses, and manual labor—these were impressive timber harvests from a single watershed. But the figures also show disturbing facts about the diminishing size of the sawlogs handled by the boom. From 1833 to 1840 the size of the sawlogs averaged 337 board feet/log, which dropped to 293 from 1841 to 1848, to 196 from 1849 to 1857, and dwindled to 66 as a twenty-year average ending in 1927.[21]

A 92-foot drop in the river from Old Town to tidewater at Eddington Head, a short distance downstream, provided the power to begin to process the timber cut in the north. What was once a pristine fishing spot along the Penobscot at Old Town now spawned sawmills.

The first was a double mill built by Richard Winslow in 1798 on the Old Town Ounegan.[22] By 1833 there were as many as sixteen sawmills located near the falls. General Veazie owned the largest sawmill, and combined with his complete acquisition of the boom and thousands of acres of timberland, he was the area's largest and best known entrepreneur.

(In June 1853 he was awarded his own town, which previously was known as Ward 7 of Bangor. Needless to say, the new municipality was named Veazie.[23])

Other area mills were located near lower Old Town on Treat and Webster islands, at upper Stillwater, and at Pushaw Village, literally dotting the mighty river's shoreline. Old Town was becoming the country's foremost lumbering center and was known as "The Gateway to the North."[24]

People began to come to Old Town seeking brighter futures. Jobs connected directly with the cutting were plentiful: river transport, sawing, and milling

This artist's drawing shows Old Town situated on the west side of the Penobscot River and contains a part of Marsh Island. A steamer churns upstream to the Madawamkeag River—the heart of the lumbering industry.
From Gleason's Pictorial Drawing Room Companion *, May 20, 1854. Courtesy of the Old Town Museum*

the lumber. Equally important were those involved in providing for the needs of the lumbermen. There were farmers, blacksmiths, liverymen, teamsters, millworkers—the employment opportunities were endless.

Immigrants came, including the Irish escaping the famine in their own land. Scots, Poles, and Italians came and supported logging in many varied ways. Canadians traveled south looking for work and a reprieve from their weather. The native Penobscot labor force was already in place and eager to work.

Although this was an era of prosperity, not everyone enjoyed the growing industry. Henry David Thoreau, already a well-respected author and naturalist who frequently visited the area, wrote:

> Within a dozen miles of Bangor, we passed through the villages of Stillwater and Oldtown, built at the falls of the Penobscot, which furnish the principal power by which the Maine woods are converted into lumber. The mills are built directly over and across the river. Here is a close jam, a hard rub, at all seasons; and then the once green tree, long since white, I need not say as the driven snow, but as a driven log, becomes lumber merely.... There were... two hundred and fifty saw-mills on the Penobscot and its tributaries above Bangor, the greater part of them in this immediate neighborhood and they sawed two hundred million of feet annually. The mission of men there seems to be, like so many busy demons, to drive the forest all out of the country, from every solitary beaver-swamp and mountainside, as soon as possible.[25]

The lumber frenzy hit its peak in the early 1800s.

By 1837 things took a drastic turn. Over-speculation, combined with leveling demand, led to a regional depression, and the economic importance of logging declined. But despite the depression's negative effects, it set the wheels of change in motion. Those who stayed were forced to turn to other industries: shoe production, textile mills, tanneries, wagon-making and boatbuilding. Many returned to farming. The logging industry, although still important, was never to return to its singular glory.

Old Town had been a recognized area within the northern political boundaries of Orono, a town eight miles south named for the famous Penobscot chieftain.[26] The area of Old Town consisted of several settlements: Stillwater, Pushaw, Great Works, Old Town, and others. By 1840 the townspeople of the Old Town area felt that they needed their own political identity separate from Orono. The Old Town district was the most populated section: 2,234 people in 284 households. The town fathers did their best to design a workable government. After several attempts, Old Town was finally validated on March 16, 1840, by an act passed by the next legislature and became its own municipality, retaining the name given it by the early settlers.

Despite this gallant start, the increase in population and the growth of industries made it difficult to transact business in the widespread villages within the town. More change in local government was the topic of discussion for many years. Finally, after complying with the necessary formalities, the area of Old Town was incorporated as a city on March 30, 1891.[27]

1 Fannie Hardy Eckstorm, *Indian Place-Names of the Penobscot Valley and the Maine Coast* (Maine: Orono Press, 1978).

2 Calloway, Colin G., *The Abenaki* (New York: Chelsea House Publishing, 1989), p. 14.

3 Eckstorm, *Indian Place-Names*, p. 38.

4 *Ibid.*, p. 31.

5 *Ibid.*, p. 2.

6 *Ibid.*, p. 2.

7 Leo Bonafanti, *Biographies and Legends of the New England Indians* (Massachusetts: Pride Publications, Inc., 1968), p. 13.

8 Calloway, *Dawnland Encounters* (Hanover: University Press of New England, 1991), p. 41.

9 Bonafanti, *Biographies*, p. 13.

10 Calloway, *Dawnland*, pp. 34-35.

11 Eckstorm, *Indian Place-Names*, p. 15.

12 Calloway, *Dawnland*, p. 8.

13 *Ibid.*

14 David Norton, Esq., *Sketches of the Town of Old Town* (Bangor: S. B. Robinson, 1881), pp. 12-13.

15 Norton, *Sketches*, p. 16.

16 A cruiser is a person who assesses the possible yield of timber stands.

17 Fannie Hardy Eckstorm, *The Penobscot Man* (Somersworth, NH: New Hampshire Publishing Company, 1972), p. 55.

18 *Ibid.*, pp. 131-32.

19 *Portland Telegram*, January 15, 1933.

20 *The Penobscot Boom*, University of Maine Studies, p. 28.

21 *Ibid.* Also Norton, *Sketches*.

22 Ounegan—a carry or portage around a waterfall.

23 Bob Cardin, *Old Veazie Railroad, 1836, One of America's First Railroads* (Bangor, Maine: Galen Cole Family Foundation, 1992), p. 5

24 "Old Town, Maine—The First 125 Years, 1840–1965," *Penobscot Times*, np.

25 Henry David Thoreau, *In the Maine Woods* (New York: Thomas Y. Crowell Company, 1961), pp. 5–6.

26 Norton, *Sketches*, pp. 12–13.

27 "Old Town, Maine—The First 125 Years, 1840–1965," *Penobscot Times*, np.

2

The Birth of an Industry

Members of the Gray family first came to Maine from Scotland in 1652. One of their descendants, Alexander Gray, was sent to Old Town in 1832 to manage a sawmill belonging to his uncle and proceeded to make his home on a farm in the settlement of Pushaw. The farm supplied most of the family's sustenance, but with the growth of lumbering, he soon became more heavily involved in that business. While operating a sawmill in the Gilman Falls area of town, he introduced his sons—Wilbur, George, and Herbert—to the lumbering business.

George, the middle son, was born in 1845 and inherited his father's proven ingenuity and strong work ethic. At the age of seventeen, he and his older brother Wilbur began selling the meat they raised and slaughtered on the farm. Their business grew quickly, and they soon opened a meat and provisions store in Old Town. Increasing demand caused them to expand and look to new markets. They began buying meat for shipment to Boston. These early experiences helped prepare George Gray for the world beyond Old Town.

Leaving Wilbur to run the market, George turned his attention to the logging business. Demand was strong enough to send crews into the woods to provide railroad ties, framing materials and shingles for houses, and cordwood. In 1872 George built what is believed to have been the first groundwood pulp mill on the Penobscot River. It was located east of Dover, forty miles northwest of Old Town. In addition to logging, Gray busied himself building homes in the growing city, and in that same year opened Gray Hardware to supply the requirements of his businesses as well as others in the township. The rear of the store was adjacent to a spur of the Bangor and Piscataquis Railroad, making it convenient for shipping and receiving.

By 1880 he had sold his interest in the pulp mill and returned to the woods to concentrate on logging. He operated several camps along with his associate, George Bowman. As they developed a market that exceeded their crews' production ability, they began purchasing from other loggers.

During this time, Herbert, the youngest of the trio of brothers, was being educated in Augusta. His major was accounting, and it seemed to fit well with the family plan. He graduated in 1875, and his business sense and global outlook were quickly apparent: he was always looking for new business challenges. His first venture was a spin-off of an existing family business. Herbert used hides from slaughtered animals at Wilbur's market to make valises. The manufacture of carbon ink was another business venture. To the Grays, practicality was foremost, and both endeavors proved profitable. Alexander Gray had guided his boys well. Each in his own right was carving his future and was respected for his ingenuity, perseverance, and business sense—and they were just beginning.

Old Town had already gained a reputation for boat building, and its bateaux were particularly well known. Bateaux, produced for loggers and river drivers, were maneuverable flat-bottomed boats with a raked bow and stern and flaring sides, which were propelled with oars and poles. They were a serious tool whose job dictated their shape. Their raked ends allowed "dry foot" access and egress from the shore, from river rocks, and from log jams. Their appearance was immediately recognizable, and their place in the working world was highly regarded.

They are light and shapely vessels, calculated for rapid and rocky streams, and to be carried over long portages on men's shoulders, from twenty to thirty feet long, and only four or four and a half wide...in order that they may slip over rocks as gently as possible. They are made very slight, only two boards to a side, commonly secured to a few light maple or other hard-wood knees, but inward are of the clearest and widest white-pine stuff, of which there is a great waste on account of their form, for the bottom is left perfectly flat, not from side to side, but from end to end. They told us that one wore out in two years, or often in a single trip, on the rocks, and sold for from fourteen to sixteen dollars.... The bateau is a sort of mongrel between the canoe and the boat, a fur-trader's boat.[1]

Old Town's reputation for boatbuilding seems to have been already established and widespread. In 1842 the papers of President Martin Van Buren mention acquiring boats at Old Town for a survey party:

The party under the direction of Professor Renwick left Portland in detachments on the 26th and 27th of August. The place of general rendezvous was fixed at Woodstock, or, failing that, at the Grand Falls of the St. John [sic River]. The commissary of the party proceeded as speedily as possible to Oldtown, in order to procure boats and engage men.[2]

While the construction of the logger's bateau was at its pinnacle in Maine, the birchbark canoe, whose con-

George Alexander Gray
Courtesy of Ruth Gray

struction methods had been known for centuries, continued its evolution along a course leading to the wood-and-canvas canoe.

The birchbark canoe was a supremely functional and beautiful watercraft. The inherent properties—strength, fragility, elasticity, and bending qualities—of the wood and bark from which it was built dictated its shape. Birchbark canoes were built right side up on a "building bed" on the ground, with stakes driven around the perimeter to designate the outside shape of the desired canoe. A roll of birchbark was placed into this "trough," a framework of thwarts and rails was positioned and fastened, ribs were bent in, and plank-

ing was placed between the ribs and the bark. The component parts were lashed together with split spruce roots, and the seams were waterproofed with a pitch-based concoction. The construction process was labor intensive, and due to the natural differences in the raw materials and handcraftsmanship, it was difficult to produce large numbers of uniform canoes of a well-liked model. It was also harder to produce new canoes because the raw materials, especially the bark, were getting more scarce.

In use, birchbark canoes were relatively short-lived, and a chief complaint was that they required constant attention. The pitch used to seal their seams tended to soften in hot sun, allowing the seams to open, and the canoes leaked. When canvas became available, it was frequently used as a patching material on birchbark canoes, and it's not hard to imagine that at some point someone put canvas over a complete but leaky birchbark canoe and found that it worked well to stop the leaks—but made the canoe considerably heavier.

The trick then was how to build a strong, lightweight canoe hull, stretch canvas over it, waterproof the canvas, and have a complete, serviceable canoe without the structural component of birchbark. Some early tries reported in the sporting literature of the 1870s were portable, collapsible, open-frame canoes with canvas covering. But ingenuity and the availability of metal fastenings created a new process, and soon wood-and-canvas canoes were being built over an upside-down form. The building form was equipped with steel bands that were the width of a rib, positioned where each rib would be. First, pre-made stems and inwales were fitted to the form. The ribs were

A Morris catalog, circa 1919. Like his canoes and boats, Morris's catalogs became more artistic over the years. His quality and craftsmanship were highly refined.
Gray Collections

· · FRIENDLY RIVALRY MAKES GOOD SPORT · ·

steamed and bent over the form in line with the metal bands and fastened to the inwales. Then planking was prepared and placed over the ribs; it was fastened by driving brass tacks through both the planking and the ribs. The tack tips were turned or "clinched" when they struck the metal bands beneath, making a strong, tight fastening. (Perhaps this technology was adapted from the cobbler's craft.) Some of the planking was also fastened to the stems before the entire hull was smoothed, removed from the form, completed, and prepared for canvassing. The hull was then placed right side up in an envelope of canvas stretched lengthwise. It was forced downwards and the canvas was stretched from side to side as well, then fastened just along the sheer, where the fastenings would be covered by the outwale. The canvas was then filled, cured, and painted—or sometimes just painted with thick paint. When the paint was dry, the outwales were fastened in place and varnished, along with the interior, the stembands were screwed on, and the canoe was complete.

The first commercial wood-canvas canoe builder was Evan (Eve) H. Gerrish, who made his home in Bangor in 1875. Prior to that, while working as a guide, he experimented using canvas on a bark canoe, according to his family history.[3] His business started with the manufacture of fishing rods and canoe paddles, but soon after in 1878, he advertised his new canoe product. By 1884 he was already producing fifty canoes a year, and his reputation was spreading. His business acumen was also growing, and he had enough foresight to exhibit one of his canoes at the New Orleans Exhibition. Gerrish's success spawned other

A Carleton canoe is displayed outside the Carleton shop in the late 1880s.
Courtesy of Harold Lacadie

A classic catalog from the E. M. White Canoe Company, circa 1918. The White Canoe Company had changed hands several times before becoming part of the Old Town Canoe Company. Many of White's catalogs and records have been lost over the years.
Gray Collections

CANOES WHITE MOTOR BOATS

builders up and down the Penobscot River Valley.[4]

Just upstream from Bangor, in the small hamlet of Veazie, Bert Morris started the Veazie Boat and Canoe Company in 1882 on the second floor of his home; he soon renamed it Morris Canoe. Both Morris and Gerrish paid close attention to detail and worked hard to make their canoes aesthetically attractive as well as functional.

Guy Carleton of Old Town, whose bateaux and birchbark canoe business had been thriving since the 1870s, now added a wood-canvas canoe and offered it to his ready market of lumbermen and guides. His factory and mill were located right at the site of the boom in Old Town on the banks of the Penobscot River. Although lacking the refinement of the Gerrish and Morris canoes, Carleton's work was respectable, following the traditions of working craft.

As the 1890s approached, more and more builders were entering the industry. One of the most notable was E. M. White. White began his business in 1885, producing canoes from his home in Pushaw, ten miles northwest of Old Town in an area now known as Gilman Falls.

In the winter of 1895, White hired two workers to help him with production. Alfred E. Wickett and George T. White were employed throughout that winter to handle the increase in demand. Wickett was just twenty-two at the time. As a teen, he and his mother had moved to Old Town from Prince Edward Island, Canada. Prior to his employment by White, census records show his occupation as a carpenter. By working with White, Wickett expanded his skills.

Business was so good that White was considering moving his business to a more central location in the city where "...strangers stopping to see his product, may examine canoes without driving to Pushaw."[5] With the help of his brother-in-law E. L. Hinckley, who provided the capital, White moved his shop to Water Street near the rail yard, which was the hub of Old Town in the spring of 1896.

Throughout the 1890s the editor of the *Old Town Enterprise*, the local paper, encouraged area businessmen to pay attention to this new industry. In the December 1894 edition, he reiterated his position by admonishing, as well as predicting:

> A canoe factory would pay well in Old Town. One that would turn out a hundred or so a month. Don't let anyone make you believe that a sale of canoes cannot be created. There are hundreds of towns and cities in the U.S. with a lake or pond close by, and the canoe could be made popular in these places as the bicycle has been where good roads exist and the luxury could be had for less money. It is the small industries that will make the future of Old Town, and the canoe business will be one of them.

His prediction was about to come true.

[1] Thoreau, *In the Maine Woods*, pp. 6–7.
[2] James D. Richardson, *A Compilation of the Messages and Papers of the Presidents 1789–1897* (Washington: Government Printing Office, 1896), p. 630.
[3] Jerry Stelmok and Rollin Thurlow, *The Wood & Canvas Canoe* (Gardiner, ME: The Harpswell Press, 1987), p. 21.
[4] *Ibid.*, p. 24.
[5] *Old Town Enterprise*, December 28, 1895.

Old Town was growing, with opportunities for anyone with an entrepreneurial spirit. The trick was to find an investment that was worthwhile. The Gray family, like so many early Old Town families, was already heavily involved in a variety of businesses in town. Alexander was at the sawmill, George in logging, Wilbur had the meat market, and brother Herbert operated his own two businesses. Farm life was still a part of their lives, and they all shared a love of horses. Horses were the main means of transportation, and were indispensable for work in the woods. Alexander had taught his boys how to buy and sell with the best of the horse traders, but more important, he had taught them to treat the animals with respect. Alexander insisted that those in his employ do the same, since their livelihoods relied on horsepower. Proper treatment of the animal was smart business. Abiel Parker Bickmore agreed with the Grays on this matter.

Abiel Bickmore was born in Old Town in January 1851. He received his early education in Old Town schools and graduated from Lee Academy. His father Oliver, a cooper, was one of a number of businessmen in town who had found work in trades peripheral to lumbering. His wares were used in the fishing industry in Gloucester, Massachusetts, and he was busy enough to require a downtown storefront. Abiel's intention was to attend college at the University of Maine, but an accident to his father required that he take over his father's cooperage. Fortunately, it was located next to the Gray Hardware store.

Bickmore's business, like the Grays', relied on horsepower: Bickmore for shipping, the Grays for

3

Shaping a Business

logging. The harnesses used on draft animals often caused an irritation which produced sores called galls. A severe case of gall could render an animal temporarily unusable, affecting business productivity. Recognizing the universal seriousness of the problem, Bickmore began looking for a cure. In a two-quart dish in his kitchen in 1884, he formulated a salve that proved to promote rapid healing of horse galls. He sold "...his first orders to Moses P. Wadleigh and Moses T. Jordan to go up river [sic] to their lumber camps."[1] Despite these meager beginnings his cure worked, unlike other treatments. He called his product Bickmore's Gall Cure.

Inevitably, the Grays were aware of the discovery and were soon to confirm its success with their own horses. With local enthusiasm and financial backing from the Grays, the Bickmore Company was established, incorporating in 1892 with Abiel Bickmore, Herbert Gray, and George Gray as officers, and with George Richardson, a close family friend, as treasurer. The trio of H. Gray, G. Gray, and Richardson proved

to be a successful team and reappears in many of the early Gray endeavors.

With the product already in demand locally, they set up production in Herbert Gray's building on Shirley Street, where he was making carbon ink. It was back to back with George Gray's hardware store on North Main Street, and a railroad spur ran between the two businesses. Herbert, as sales agent, immediately began to introduce the product from the East Coast to the Midwest, traveling by horse and wagon. When his other enterprises required that he stay closer to home, he was replaced in 1896 by another family member, Wilbur's son Jesse, who acted as the company's sales representative, traveling by rail as far as California to spread the news of the great discovery. Orders from throughout the country came pouring in to Old Town, and the factory workers, mostly women, found it necessary to work evenings to supply the growing mountain of orders. By 1896 the factory produced 1,700 boxes a day of Bickmore's Gall Cure and could not keep up with demand. By the turn of the century, Jesse, George, and Herbert Gray, along with Pearl Cunningham, a salesman from the hardware store, all found themselves on the road marketing the product. But the Gall Cure didn't stop there. News had already reached Canada, and foreign orders were coming in as quickly from the European countries. Bickmore's Gall Cure was just short of a miracle!

With so many things on the Grays' business plate, it's hard to believe they could find time for more. Bickmore's Gall Cure was thriving, the logging business was demanding, the hardware store filled a necessary niche and certainly supported the other businesses with tools and equipment, and Herb's valise business did well for its time. The carbon ink factory was not as successful as Herbert had hoped and was eventually sold to a company in New York.

As visionaries, the Grays began to turn their attention in the late 1890s to the manufacture of canoes. It was an industry waiting to happen, just as the editor of the *Old Town Enterprise* prophesied in 1894.

The Grays did not intend to become hands-on canoe builders. Although family members were an important part of the Gray workforce, their forté was management. They needed to find an operations manager who knew the product and had technical know-how. Alfred E. Wickett seemed to be the right person.

Wickett was a young, ambitious, experienced carpenter and canoe builder who had worked for E. M. White as early as 1895 and may have operated his own canoe-building business for a short time before being hired by George Gray. Local legend has it that a shop was set up in a back shed of the Gray Hardware store, and that a few canoes were built to test the product and the market in the late 1890s. With Gray supporting Wickett's ambition, a company was born that would become a giant in the industry.

In the year 1900, four hundred canoes had been manufactured in Old Town, and Wickett's work was included in this number. Sales were brisk and the Gray operation had to be moved to more appropriate surroundings. On October 13, 1900, the *Old Town Enterprise* announced the addition of another factory to the town's busy industries. A canoe plant would be opening, "utilizing the second and third floors of the

The only catalog known to be issued by the INDIAN OLD TOWN CANOE COMPANY, circa 1901.
Gray Collections

Sailing canoes continued to be popular from the early years on. This American Canoe Association regatta was held at Seminole Canoe Club in Jacksonville, Florida, probably in the early '20s. Gray Collections

*An INDIAN OLD TOWN CANOE
COMPANY leaflet describes some of
their wares.*
Gray Collections

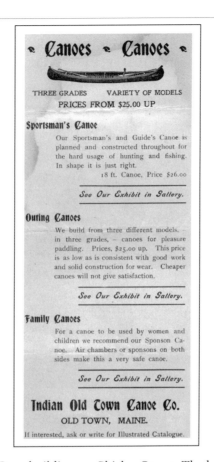

building savvy was soon apparent, and several models were offered. In an attractive catalog, INDIAN OLD TOWN introduced a line of canoes, including sailing models, a rowboat, and accessories.

The new company offered five different canoe models. The premier canoe was the HW model. Through the years the canoe was mistakenly called the "Henry Wickett" canoe, thought to be named for the builder—but Wickett's first name was Alfred. Perhaps he may have named the HW after his father Henry as a tribute, but more likely its name is attributed to its design. With a more rounded bottom than that of other models, the HW was intended for "heavy waters," as the sales literature attested. The rounded bottom offered greater overall stability in rougher conditions; the bow and stern were full, giving it extremely good riding qualities and buoyancy in rough water or ocean swells. (A canoe with a rounded bottom is not as initially stable in calm water as a flat-bottomed canoe, but it is more agile in rough conditions.) In many cases, novice paddlers ordered the HW with sponsons. These sponsons were long, hollow air chambers that were placed high on the sides of the canoe extending from stem to stern. Amidships they measured about four inches wide and tapered toward each end. They only came into play when the canoe was leaned over, and their added width and extra buoyancy made it extremely difficult to capsize the canoe. A circa 1901 INDIAN OLD TOWN CANOE COMPANY catalog stated:

A sponson canoe is very hard to upset. It is particularly fitted for all who wish a paddling, rowing, and sailing canoe combined.

Herbert Gray building on Shirley Street. The building was designed for manufacturing, with long rows of windows surrounding the spacious floors. There will be ample light, and the powerful engines used in the manufacture of the Gall Cure on the floor below will supply extra power and heat for all necessary requirements in that direction."

Recognizing the important bond between the canoe and the Indians, the new venture was aptly called the INDIAN OLD TOWN CANOE COMPANY. Wickett's boat-

In 1970 I came into possession of what was thought to be an 18-foot OLD TOWN canoe. In those days, I was interested in running rivers, etc., rather than in old canoes. My neighbor, Babe, and I stripped the old canvas off the canoe and recovered it with fiberglass. We didn't do a very good job and often joked that its finish looked like it had been completed using the residue of giant seagulls, on the wing. Our goal was to enter the First Annual Memorial Day River Rat Race on the Blackstone River in Woonsocket, Rhode Island.

This was the first time we had participated in an event of this type. We cartopped our ugly canoe on my 1968 Chevelle wagon to the starting point, which was at the old wading place below the falls. In the late 1600s and early 1700s, Indians and early settlers passing through the area from the Massachusetts Bay Colony had crossed the river here on their way to the Connecticut Colony. The Indians named the beautiful area "Woon Soket." Translated, it means "thunder mist," which clearly describes the water after it passes over the natural rocky falls into the gorge below.

We lined up along the shore, waiting to be called to the starting point. Every canoe would be timed. We were given the starting flag, and with Babe in the bow, began paddling as if we were trying to escape an oncoming lava flow! Our wives and kids cheered us on, and we no sooner entered the first rapids than we hit a submerged rock. Babe was tossed out of the boat but got back in, but we soon realized that we were taking on water. A quick look under the life preserver Babe was kneeling on revealed a large hole. Babe kept the preserver directly over the hole and we continued our mad paddling.

The river soon widened and became deeper. Water was seeping into the canoe and before long it was half full! Boy, did that big, old canoe cut through the water. We finally reached the finish line in Cumberland, five or six miles downriver, and everyone ran to our aid, fearing that at any second the boat would disappear. Ten grown men pulled the canoe from the river and drained it.

When the last canoe's time was announced, we were awarded a respectable third prize, considering we had "rocked" the canoe. It had performed admirably despite us.

I developed an attachment to that relic and had it restored to its original condition. My wife shares in my enjoyment of pad-dling the pod, now gleaming with a coat of green paint. The River Rat Race was just the beginning of a twenty-eight-year friendship with a piece of history that I've paddled many more miles on many ponds, lakes, and rivers.

—Harvey Greenhalgh, Jr., Harrisville, RI

When we are paddling our restored 1926 OLD TOWN HW, we are often amazed as we paddle by groups of people on shore. Almost invariably, the first to acknowledge our presence are the younger kids. They incessantly pester their parents and insist on showing them the "pretty canoe." Leave it to the kids!

—T.L.B., Muamee, OH

Later catalog pictures supported these claims by having three or more people sit on the gunwale of the canoe,[2] yet it was still upright and afloat. In an age when many people could not swim, sponsons were very reassuring and customer response was positive:

"...purchased one of your 17-foot sponson canoes. I used, and allowed my children to use this canoe all the season...(sometimes in quite rough weather) and am so well pleased with it that I have sent an order for another of the same kind and size to be delivered in the spring...." —E. R. Brown, President of the Strafford Savings Bank, Dover, NH, December 31, 1913

With options like the sponson and strong regard for customer satisfaction, it is no wonder that INDIAN OLD TOWN CANOE COMPANY sales were increasing.

Another popular model was the IF model, advertised as "a special Indian model...planned by one of our Indian workmen."[4] It was built for quickwater to meet the requirements of the increasing number of

THE INDIAN OLD TOWN CANOE CO.

Are builders of canoes for all purposes and of various grades. Our models are planned to meet the requirements of all. Our workmen have had years of experience. We use the best of materials. Every canoe is thoroughly put together and guaranteed strictly as represented. Our canvas canoe is all that Indian invention and knowledge and American workmanship and aim for improvement can make of a canoe. It is impossible to show the fine lines of a canoe in a small picture.

MODELS

H. W. Model

This model is particularly adapted for the man who spends his leisure hours upon the water, loving the recreation and gaining health and strength from the most exhilarating sport of canoeing.

This canoe is graceful in shape — is rather full at bow and stern giving it particularly good riding qualities and making it very buoyant in heavy seas. It is steady and will work well in salt water. We particularly recommend it.

guides and sportsmen. IF, which stood for Ideal for Fishing, was a model designed with generous beam, a flat bottom, and good load-carrying qualities. The bow and stern narrowed sharply, and the 1901 catalog described it as a quick canoe, good for river, still pond, or poling over rapids. This canoe boasted the ability to draw very little water, making it effective in the shallows and rapids.

A canoe designed for commercial use was the Guide's Special, tailor-made for the professional guide. It featured planking in long lengths, which made the bottom stronger, and a stern seat mounted high enough to afford a view over the "sport" or client. It had decks, thwarts, and stems made of white ash, and spruce inwales and outwales. Its stern seat was canvas or caned; professional guides often paddled while their client sat on the bottom of the canoe in a low, folding chair. When desired, a bow seat was furnished at no extra cost.

The canvas on the Guide's Special was a heavy number six canvas. (In modern times, many builders choose to use a lighter number ten canvas.) The Guide's Special was a utilitarian boat in every sense: "no money is put into ornamenting."[5] It was offered in 18-, 19-, and 20-foot models, with the latter described as perfect for salmon fishing. Prices ranged from $26–$30.

INDIAN OLD TOWN Canoes could be purchased in three grades: AA, BB, and CS. The double A grade was the most expensive and offered several enhancements to the discriminating owner:

Ribs and planking of selected Cedar. Stem and

stern posts of White Ash or Oak. Gunwales of best Spruce. Outwales of selected Spruce. Decks, Thwarts and Seat Frames of Mahogany, Birds Eye Maple or Quartered Oak. Seats Caned or Upholstered in Chase Leather. Bang plates of polished brass. Canvas of first quality. Copper and brass fastened.

Painting and Finish: In this grade special attention is given to the finish. The canvas is carefully filled and rubbed down, painted and varnished. The wood work is filled, varnished and rubbed down with pumice stone in oil to a smooth dull finish. Will be left bright if desired.

A second grade that only shows up in the 1901 catalog was the BB, a mid-priced alternative:

Ribs and Planking of Cedar free from knots. Stem and stern posts of White Ash. gunwales of Spruce. Outwales of Spruce. Decks, Thwarts and Seat-frames of Birch, Maple or Oak. Seats caned. Brass bang plates. Canvas of first quality. Copper and Brass fastened.

The painting and finish in this grade are good throughout. The canvas is carefully filled and rubbed down, well painted and varnished. The interior is brought to a smooth surface and varnished.

The more utilitarian-minded buyer might have preferred the CS grade. The CS distinction indicated that the canoe was made of cedar and spruce. More practically, however, the letters stood for Common Sense. The Grays intended this grade for the common man,

With much elbow grease, an old war horse can be restored to a beautiful 18-foot OLD TOWN Guide model once again. Carl Klem Photo

The Double End Boat was first introduced in 1901 as a Canvas Covered Rowboat. This 1945 model has been beautifully restored by Rowland Thompson of Olympia, Washington.
Courtesy of Patrick Chapman

someone with a no-nonsense approach to canoeing—someone, perhaps, with limits on his wallet.

This grade of canoe we build for buyers who wish to pay a moderate price for a good serviceable canoe, well built, out of good materials. No shaky or unsound lumber is used in this grade. This is a common sense canoe, is stiff, and will wear well.

If only equipment were still available at these prices! INDIAN OLD TOWN CANOE COMPANY catalog, circa 1901. Gray Collections

Miscellaneous Price List

Paddles	Spruce, finished in varnish.			$1.00 each
	Ash	"	"	1.25 "
"	Maple	"	"	1.50 "
"	Birds Eye Maple finished in varnish,		1.75 to 2.25 "
"	Decorated Indian designs in pyrography, Maple and Spruce,		2.75 and up
"	Decorated Indian designs in pyrography, Birds Eye Maple,		4.00 and up
Oars	Spruce, per foot,			.10 to .15
Chairs	Folding, slat,			1.00 each
Back Rests	Plain board, varnished,			.35 "
" "	Slat, light and good,			.50 "
" "	Hard wood frame, caned, an excellent back,		1.00 "
" "	Upholstered awning stripe, cork filled,		1.00 "
" "	" " " moss "		1.25 "
" "	Chase leather or corduroy— hair filled,		2.25 "
Cushions	Awning stripe — cork filled — 14 x 14 x 2 1-2,		.75 "
"	" " " 14 x 24 x 2 1-2,		1.00 "
"	" " — moss filled — 14 x 14 x 2 1-2,		1.00 "
"	" " " 14 x 24 x 2 1-2,		1.50 "
"	Chase Leather or Plain Corduroy — hair filled, to match canoe, to order only 14 x 14 x 2 1-2,		2.00 "
"	Chase Leather or Plain Corduroy — hair filled, to match canoe, to order only, 14 x 24 x 2 1-2,		3.50 "
Pillows	Cork, Down or Feather covered to order.		
"	Fir Balsam, — plain, to be covered,		1.00 "
"	" " " covered to order,		
Carpets cut to fit bottom of canoe and bound, to order.			
Rowlocks	Fancy side plate Galvanized Iron No. 0. $1.00, No. 1, $1.25 pair		
"	" " " Polished Brass, " 1.75, " 2.00 "
"	Outrigger — Galvanized Iron,		2.00 "
"	" Polished Brass,		6.00 "
Flag-pole sockets,			.75 each

Prices quoted upon other fittings on request.

Ribs and Planking are of Cedar: Gunwales and Outwales of Spruce. Posts, Decks, Thwarts, and Seat-frames are of White Ash. Seats caned. Brass bang plates. Canvas of high grade. Copper and brass fastened. This grade is painted and finished to give good service. Has dull finish outside, natural finish inside.

In an attempt to round out their line to appeal to everyone, INDIAN OLD TOWN introduced the FB model. The FB was only offered in a 17-foot length but was available in any of the grades, or quality options, of the time. This model had a flat bottom, with the name FB indicating that shape. It was advertised as a canoe for general work for those who wanted the feeling of initial security offered by a flat-bottomed canoe.

Another utilitarian model was the GG, perhaps an abbreviation for "good general." Its hull design was similar to that of the IF model but was not as sharp in the bow and stern. The GG canoe was also "beamier," 35 inches compared to the IF's 34 inches. This model was only produced in an 18-foot length and sold from $26–$41, depending on the grade. The GG was a compromise between the rounder bottom of the HW and flatter bottom of the IF.

In addition to the paddling canoes, sponson sailing canoes could be ordered and were built to suit the buyer. In 1901 a canvas-covered rowboat, later to be called the "Double End Boat," was introduced. The catalog boasted its advantages over the "ordinary heavily constructed rowboats" of the time, claiming that INDIAN OLD TOWN rowboats were light, water-tight, and required "no soaking to swell planking." Heavier ribs, planking, and canvas were used on the rowboats than on the canoes. Three sizes were offered: 14, 16, and 18 feet. Each length was offered in two grades: AA and BB. Prices began at $40 in the AA grade and $35 in the BB grade for the 14-footer. The price went up by $5 with each 2-foot increase in length.

Most models offered all grades, and sponsons could be secured to any model for $10. The fledgling company wanted to be sure its customers had a wide assortment of accessories as well. The inventory varied from paddles to flag sockets and everything else paddlers might need. Paint color choices included dark red, bright red, dark green, and light green. Any special color desired could be furnished to order. Personal decoration was encouraged. Attention to detail and a personal touch could also be added for a cost. Names could be added to a canoe for twelve cents a letter, a gold stripe for $2.00. In addition to adorning your canoe, you could also choose a paddle with an artistic motif. Paddles decorated with an Indian design through pyrography (wood burning) were available starting at $2.75, depending on the wood used and the intricacies of the work.

In its first year, the INDIAN OLD TOWN CANOE COMPANY boasted a productive output of some 250 canoes.[6] The sales lived up to every expectation. The company's advertising was minimal in the early years; in 1900 it was limited to an expenditure of only about $100, with ads running in sporting and recreation magazines and papers read by perspective buyers.[7] *Forest and Stream*, the predecessor of the modern-day

Field and Stream, was a common venue. The demand was so great that the company's quarters on Shirley Street could no longer meet its needs, and it expanded, moving into a large, four-story wood frame building on Middle Street, formerly the Keith Shoe Factory, that seemed ideal for upcoming operations. The canoe company retained the two lower floors and basement of the structure, and the upper floor was used for production of Bickmore's Gall Cure. The Shirley Street building was used as the finishing department and for shipping and storage.

This unique segmented canoe was built in 1901 for Baldwin Ziegler's expedition to the North Pole; it was designed to be shipped easily on dog sleds and reassembled at its destination. In the photo, it's shown in front of the former Keith Shoe Factory on Middle Street, the new home of the INDIAN OLD TOWN CANOE COMPANY.
Gray Collections

1   "A. Parker Bickmore: Originator of Bickmore's Gall Cure," *Old Town Enterprise*, March 5, 1901, p. 1.
2   A gunwale is a side rail of a canoe. Inwale and outwale refer to the inside or outside rails or gunwales.
3   OLD TOWN CANOE COMPANY catalog insert, 1903.
4   INDIAN OLD TOWN CANOE COMPANY catalog, circa 1901.
5   *Ibid.*
6   *Old Town Enterprise*, March 4, 1905.
7   S. B. Gray, "How We Built New Markets for an Old Product," *System, the Magazine of Business*, January 1927, p. 56.

4

The Outdoor Life

The first buyers of INDIAN OLD TOWN canoes were often the men who used them to earn a living, and INDIAN OLD TOWN had a ready market nearby.

The groundswell of men flocking to the woods, which started in the late 1800s, was continuing to grow throughout the country. One exodus lead city dwellers to the Adirondack Mountains of New York State. Thousands were drawn by the healing powers of the woods promised by a preacher named William H. H. Murray, whose book, *Adventure in the Wilderness*, written in 1869, portrayed the area as a medicinal Mecca. Murray was credited with having "kindled a thousand campfires and taught a thousand pens how to write of nature."[1] He was certainly not the first to promote the curative qualities of outdoor life, but he was one of the first to successfully record his perceptions in such a way that anyone could understand and relate to his images.

Taking the bow seat, John paddled straight for the west shore of the lake, and the light boat, cutting its way through the lily-pads, shot into a narrow aperture overhung with bushes and tangled grass, and I saw a sight I never shall forget. We had entered the inlet of the lake, a stream some twenty feet in width, whose waters were dark and sluggish. The setting sun yet poured its radiance through the overhanging pines, flecking the tide with crimson patches and crossing it here and there with golden lanes. Up this stream, flecked with gold and bordered with lilies as far as the eye could reach, the air was literally full of jumping trout.[2]

Tourists flocked to the woods in any way they could. Once there, they used Adirondack guideboats, rowboats, and canoes to seek the adventure portrayed in Murray's written pages. New builders began to emerge, the most notable being J. H. Rushton of Canton, New York, who began building in 1873. His boats were sought after for his high-quality workmanship, and his marketing abilities were unparalleled for the time. Genius like Rushton's did not go unnoticed. His success encouraged a number of others, including John Ralph Robertson.

John Ralph Robertson was born in Canton (now reverently called "Rushton Country"), and although there is no definitive link between Rushton and Robertson, he was certainly affected by Rushton.

Other members of the Robertson family were already established in the Lawrence, Massachusetts, area and as a young man, J. R. Robertson left Canton and joined them working in the woolen mills. But indoor work of this type did not seem to suit him, and in 1881 he joined efforts with a man named Holmes, opening a boatbuilding shop on the banks of the Merrimack River. Their circa 1884 catalog advertises their products as "ADIRONDACK PORTABLE SPORTING BOATS AND CANOES...," and the title page inside the catalog clearly states the Rushton link: "ADIRONDACK BOATS AND CANOES...(KNOWN AS RUSHTON'S PORTABLES) FOR HUNTING, FISHING, TRAPPING, ETC. These boats and canoes are designed by J. R. Robertson formerly of Canton, St. Lawrence County, New York."

Robertson virtually copied Rushton's catalog and used it as his own! He replicated Rushton's designs—

In second canoe, foreground, William F. Kip in stern seat, then foreman of the J. Henry Rushton Boat Shop in Canton, New York.
Courtesy of the Adirondack Museum

An Indian guide takes a moment out to pose for a picture with his client on the Penobscot River. Guides often used long paddles that enabled them to stand and check conditions as they paddled downstream. Indian Island is in the background.
Gray Collections

the Stella Maris, the Grayling, the Princess, and others—but was never able to equal Rushton's building mastery nor, apparently, did he try. Certainly the patent and copyright laws were not as stringent then as they are today, but Robertson's tactics were clearly questionable. Rushton himself criticized builders like Robertson in a letter included in the introduction of his 1885 catalog:

> We are pioneers in light boat building and an increasing trade with each succeeding year shows that we are appreciated by our patrons. This success is all the more gratifying to us, as it comes in spite of competitors, some of whom are none too particular what representations they make to the public if they only sell their goods. In one case these were such as to lead strangers to believe we were connected with another firm and that they would get the very same goods from either. Right here we wish to say that we are connected with no other firm any where, and that no other firm builds boats or canoes identical with ours whatever they may call them.[3]

The demand for pleasure craft spread throughout the New England area as converts were added to the sport. The Penobscot River basin lured the sportsman, and in Maine many men were turning to guiding to fulfill the demand.

> He who continually lives along the coast line knoweth not the benefit of the North Woods. The word humidity is not in the bright lexicon of the Pine Tree State. Poor appetite has no abode there. Indigestion is a stranger in the land. Stomachs that rebelled are forgiven and forgotten. Nervousness soon seeks other climes. Imaginary evils vanish into thin air. What seems mountains elsewhere to the tired brain become molehills. Morbid thoughts give way to pleasant reflection. The inward antipathy hidden by outward courtesy of man towards man resolves itself into the true Christian spirit. Woman's jealousy of woman has no abiding place in the woods of Maine. The struggle for worldly goods that is driving many business men to the asylum and penitentiary ceases for the time being. In

Like their counterparts in the Adirondacks, men of the Cherry Valley Group at Gravelly Point, Ostego Lake (source of the Susquehanna River, near Cooperstown, New York), gather around a very early OLD TOWN canoe for a camp shot, circa 1910.
New York State Historical Society Assoc., Cooperstown, NY

J. Henry Rushton, Henry Rushton,
Jim O'Brien, and Dr. Mark Henley
are shown at North Branch Camp.
Courtesy of the Adirondack Museum

fact, the surroundings there give one that quiet repose that enables you to see this life as it should be seen.[4]

Maine, which offered a vast array of game to hunt, rivers and lakes to fish, and mountains to climb, became increasingly more tourist oriented. According to the local paper, The *Old Town Enterprise*, by 1902 there were already 1,802 guides registered in the state. They escorted 4,124 residents and 9,199 non-residents into the Maine woods. The INDIAN AND OLD TOWN CANOE COMPANY helped answer the call of those who wished to experience the wilderness.

[1] Warder H. Cadbury, "Introduction to Adventures in the Wilderness," p. 12. Quoted in Murray, "Reminiscence of My Literary and Outdoor Life," *The Independent* (New York) LVII (1904), p. 278.

[2] William H. H. Murray, *Adventures in the Wilderness* (New York: Syracuse University Press, 1989), p. 67.

[3] Dave Baker Collection.

[4] G. Smith Staton, *Where the Sportsman Loves to Linger* (New York: J. S. Ogilvie Publishing Company, 1905), p. 34.

5

The Canoe Craze

While the Penobscot River was often clogged with logs waiting for processing, further south, around Boston, another river was equally crowded—with canoes. It was said that on warm summer evenings and weekends, one could have crossed the Charles River by stepping from one canoe to the next. The popularity of canoeing quickly grew beyond anyone's expectations. This newer, lighter craft was likely to be a young man's first major purchase, just as the automobile would be to the next generation. A canoe combined athletic prowess and exercise with a means of transportation. It also provided a method to "break the ice" with members of the opposite sex. "Would you like to go for a spin in my canoe?" was the classic introduction that turned many a young girl's head. If one didn't own a canoe, one be could rented at a local boathouse for twenty cents an hour.

More than a dozen public boathouses cropped up along the Lakes District of the Charles, in the towns of Newton, Waltham, and Weston, the area known to the locals as "Riverside." The boathouses provided more than just canoes; they offered their patrons social halls with dance floors and music rooms, game tables, bowling alleys, restaurants, and verandah seating with a view of the river. On Saturday nights, boathouse owners further encouraged river activities by having a motorboat pull a scow up and down the river with two or three bands on board who played for the paddlers' entertainment. Canoes, as many as two hundred, would "raft up" behind the scow and join in, adding their voices to the melodic tones.[1] Although few of the boathouses remain today, it is still easy to see why this place, with its picturesque river and well-shaded paths, became so popular.

Affiliated with each boathouse were canoe clubs that sponsored sporting events, parties, band concerts, river carnivals, and fireworks:

> Instantly the river basin gleamed in every direction with red fire; flaming balloons soared into the sky, rockets burst on high, showering from the heavens as falling stars, and explosions boomed all along the shore.... For forty minutes the pyrotechnics and the music continued before the flotilla emerged into view from under the Prospect Street Bridge into the mile of curving river designated as the chief course of the parade. It is estimated that there were fully four hundred illuminated canoes in line and in addition there were many floats. Beside these there were several thousand canoes and other pleasure craft on the river....[2]

These clubs became the focus of recreational activity in the Boston suburbs, and each club vied for the public's attention. At the Boston Athletic Association boathouse in Weston in the winter of 1896–1897, the BAA planning committee came up with the idea for a marathon race to kick off the spring season. This was the birth of the Boston Marathon.[3]

Along with its numerous boathouses, the area offered another recreational opportunity. The Commonwealth Avenue Street Railway, a local traction company, built a large amusement park along the riverbank in the Auburndale section of Newton.[4] Norumbega Park, which opened in 1897, drew tens of thousands of revelers to the riverside each summer weekend. Initially ten acres in size, the park eventually expanded to twenty-five acres. Created as a means of increasing patronage on the trolley cars, Norumbega became synonymous with excitement and quickly turned into one of the most popular recreation sites in New England. The park was bordered on two sides by the Charles River and on a third side by busy Commonwealth Avenue. The electric cars of the street railway brought throngs of city-dwellers searching for a reprieve from Boston's summer heat. Although Norumbega Park was only one of a score of riverside attractions, fully one-third of the trolley patrons visited the park. Norumbega's huge outdoor theater

Scene from annual field day in water sports at the Boston Athletic Association Boat House on the Charles River, circa 1900.
Courtesy of Robert Pollock

presented a vaudeville bill that changed weekly, motion pictures, and concerts. There was also a world-class restaurant, a Dentzel carousel, a penny arcade, midway games, picnic areas, and an electric fountain. The park's zoo was the largest in New England, with a diverse assortment of animals, including "Mountain Chief, the biggest bison in the world, weighing in at 2,500 pounds."[5] In addition, Norumbega boasted acres of flowery vistas and scenic glens. Trolley travelers not interested in Norumbega headed for the river and a canoe.[6]

The amusement park also advertised the largest canoe livery in the United States. Its two boathouses had storage space for six hundred canoes, half of which was reserved for its own livery fleet, with the balance rented out to house private canoes for $10 to $12 per canoe a year. The park was situated at the midpoint of the six-mile-long Lakes District, bounded upriver by the waterfalls at Newton Lower Falls and downriver by the Boston Manufacturing Company's

dam at Moody Street in Waltham. The *Boston Globe* of May 6, 1902, quoted Superintendent Albert Habberly of the Metropolitan Park Commission, speaking about the Charles between Newton Upper Falls and Waltham: "More canoeing is done on this stretch of river than is done on all the rivers in the rest of the state of Massachusetts, put together."

With this rapidly expanding interest in canoeing, problems were inevitable. People living along the river became particularly incensed over the large numbers of couples who were paddling their canoes into coves and bays for romantic encounters. In response to complaints from residents of Newton and Waltham, rules for boating etiquette and safety were established by the Metropolitan Park Commission in 1903:

Rules and Regulations for the Government and Use of the Waters of Charles River

Rule 1. No person shall annoy another person, or utter any profane, threatening, or abusive language,

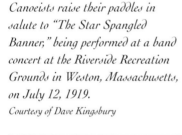

Canoeists raise their paddles in salute to "The Star Spangled Banner," being performed at a band concert at the Riverside Recreation Grounds in Weston, Massachusetts, on July 12, 1919.
Courtesy of Dave Kingsbury

or loud outcry, or solicit any subscription or contribution; or have possession of or drink any intoxicating liquor; or play any game of chance; or have possession of any instrument of gambling; or do any obscene or indecent act; or have possession of or use a flash light, search light, or dark lantern.

Rule 2. No person shall throw any stone or other missile, or have possession of any firecracker, torpedo or fireworks; except with written authority from said Metropolitan Park Commission, engage in business, sell or give away any goods, wares or circulars; or post, paint, affix or display any sign, notice or placard or advertising device; or have possession of or discharge any firearm.

Rule 3. No person shall bathe except in proper costume and a place designated therefore.

Rule 4. No person shall have charge of, run or drive a boat propelled by a steam, naphtha, gasoline, electric, or other motor or engine, unless he shall first have obtained a permit therefore from the Metropolitan Park Commission.

Rule 5. No person shall run or drive a boat propelled by a steam, naphtha, gasoline, electric, or other motor or engine at a speed exceeding eight miles an hour; or in such a manner as to endanger or annoy the occupants of other boats or canoes.

Rule 6. No person shall row or paddle a boat or canoe unless able to control or handle the same with safety to himself or to other occupants thereof; or in such a manner as to annoy or endanger the occupants of other boats or canoes.

Rule 7. No person shall throw, drop or place in the river any waste paper, rubbish or refuse; or, except with written authority from said Metropolitan Park Commission, moor any boat or raft in the river or build or maintain any float or platform on or over

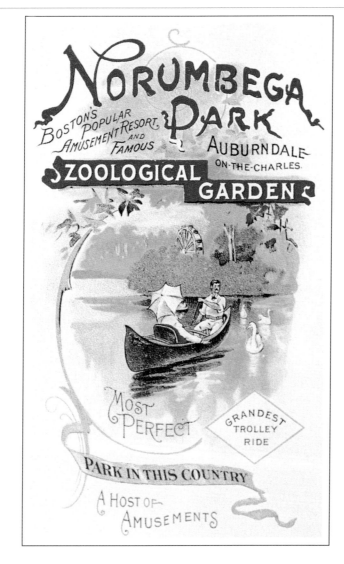

NORUMBEGA PARK

BOSTON'S POPULAR AMUSEMENT RESORT, AND FAMOUS

AUBURNDALE ON-THE-CHARLES.

ZOOLOGICAL GARDEN

MOST PERFECT

GRANDEST TROLLEY RIDE

PARK IN THIS COUNTRY

A HOST OF AMUSEMENTS

The younger members of the Dayton Canoe Club often try to coax the older guys to tell us tales of canoeing adventures in the past. Sometimes this leads to a lot of tall tales, and we end up rolling on the floor, laughing.

The discussion once got around to courting in a canoe. An old-timer began by comparing the comforts and atmosphere of a canoe to the horsehair seats of a Model A Ford, adding that the cost was too high for most young men to own a car. He described how they would take all the cushions off the front porch furniture to line their floating love nests—not always from their own front porches! He actually claimed that the song "A Hot Time in the Old Town Tonight" referred to sparking in a canoe!

—Bill Whalen, Dayton, OH

Any person violating any one of the above rules is liable to a fine of twenty dollars for each offense.

The rules paint a picture of a wild and rowdy time, and they may have been justified. But they caused a furor of indignation from the canoe club members, boathouse owners, and others involved in the sport. They felt the rules were uncalled for and also unenforceable. The phrase "indecent act" was construed to prohibit couples kissing in canoes or anywhere along the river. The *Boston Herald* remarked on August 24, 1903:

> It may not be wicked to go canoeing on the Charles with young women on Sunday, but we continue to be reminded that it is frequently perilous.... The canoeist arrested for kissing his sweetheart at Riverside was fined $20.00. At that rate it is estimated that over a million dollars' worth of kisses are exchanged at that popular canoeing resort every fine Saturday night and Sunday.

the water of the river; or fish from any bridge which crosses the river; or cut or injure the ice on the waters of the river in such a manner as to interfere with the lawful rights of the public thereon.

The protests continued through the spring of 1903, prompting editorial cartoons poking fun at the new rules: "All the world loves a lover, except at horrid Riverside on the Charles." Graffiti featured "Farewell to Romance, here comes a cop...," and stories of the protest appeared in newspapers across the United States. Even popular song lyrics reflected the sentiment, "I asked her for a little kiss...down by the Riverside, down by the Riverside, down by the Riverside...."

Petitions were sent to the governor voicing the public outcry. But the boating zealots didn't stop there. Police officers who attempted to enforce the new rules were subject to verbal abuse and even sabotage, as angry canoeists struck back. One interesting protest involved the positioning of two dummies in a canoe in what might be called a very compromising position. The pranksters would then tether the canoe and hide themselves in the nearby woods. When the officers paddled up and attempted to reprimand the "law-breakers," they were met with catcalls and rude remarks from the riverbank.[7]

For decades, the Charles River was a center of recreational activity and the canoe was at its core. Shortly after the turn of the century, the Metropolitan Park Commission estimated that more than 5,000 canoes were berthed along the Lakes District, and by 1920 that number had grown to 6,000.[8]

Canoeists protested their displeasure toward regulations aimed at controlling their behavior on the Charles River.
From the *Boston Post, August 19, 1903.*

[1] Arthur F. Joy, " A Canoe, the Charles River, and You," *Yankee Magazine*, October 1957, p. 54.

[2] Ralph Frederick Perry, *Canoeing the Charles: Images and Field Notes from 1902-12* (Hollis, NH: Hollis Publishing Company, 1996), p. 63.

[3] Robert Pollock, *Out to Norumbega*, unpublished manuscript, 1996.

[4] "Traction" company was a common and accepted way to refer to street railways or trolleys a century ago.

[5] Pollock, *Out to Norumbega*, np.

[6] Joy, *Yankee Magazine*, p. 52.

[7] Pollock, *Out to Norumbega*, np.

[8] *Ibid.*

6

Meeting Market Needs

Boston's passion for canoeing also affected the INDIAN OLD TOWN CANOE COMPANY. The Bickmore salesmen, who already garnered orders for the miracle salve, doubled their efforts by taking orders for canoes from the dealers they had established along their routes: hardware stores, retail shops, and livery areas. Back at the factory, orders were increasing and A. E. Wickett's younger brother Richard joined the force.

The numbers were good enough for George and Herbert Gray to take a hard look at their budding business. They were not manufacturers and designers themselves, and that fact made them unique. Most of the builders at the time attempted to do it all, but George Gray set out to find the best people and put them in key positions.

Gray's next move was a daring one. Unaware of the earlier ethical cloudiness, he hired John Ralph Robertson, now in Auburndale, Massachusetts. Drawing originally from Rushton's work, Robertson had established a reputation on the Charles River as a builder of wood-and-canvas canoes and was a key player in the marketplace. He offered manufacturing knowledge and a market outlet on the Charles River for Gray's canoes. Although the Holmes Robertson Company of Lawrence had been successful, the lure of the Boston-Charles River market was great, and he had set up a very successful shop in "Riverside" near Norumbega. Now he was connecting with Maine.

Mr. J. R. Robertson is a canoe and boat builder of twenty years experience, well-known among boating people. Thousands have used his canoes on the

Charles River and appreciate the graceful design, thorough workmanship, ease of handling, and general good finish...he has brought the filling and painting of canvas canoes to a very high standard.[1]

The Grays decided it was time to incorporate, and in January 1902 the articles of association were finalized. From the association's log:

Purpose of manufacturing and selling at wholesale and retail, canoes and boats, and equipments therefore of all kinds and description, owning, buying and selling, leasing, and hiring real estate and personal property necessary or convenient for the sale and manufacture of aforesaid articles and using and performing all things necessary or convenient for the carrying on of aforesaid business. Also for the purpose of manufacturing, trading, and merchandising.

The articles were signed Herbert Gray, George Gray, George Richardson, and John R. Robertson (Auburndale). Each man received a copy of the Articles of the Association on January 16, 1902, as testified by Herbert Gray's notes in the Association's log. The actual papers, filed with the secretary of state, were signed on September 23, 1902, by the directors: Herbert Gray, president; George Gray; George H. Richardson, treasurer-clerk; and John R.

From the beginning, the Gray family hired some of the best. This yacht dinghy was designed by B. B. Crowninshield, who was once described as a "well-connected Boston yacht designer with blueblood ancestry seeming to go back to God Himself."
Gray Collections

In 1945 our first child had just been born and we had bought our first house in Milford, Connecticut, a small town on Long Island Sound. I had little money for luxuries but had managed to find a local man who built rowboats: square, boxy, flat-bottomed craft which he sold for $50. I bought one and tied it up at the town dock. Captain Botsford, a descendant of the earliest settlers of Milford and the only user of the town dock, operated a small fishing boat on a daily basis. Some of us would go down to the dock in the evening when Captain Botsford returned from his dragging to purchase flounder and blackfish. One evening, after getting my fish, I noticed a beautiful new skiff with a varnished transom and rails, tied up alongside. I bailed out my boxy rowboat and climbed back to gaze at the skiff. The skiff's owner turned up, and I commented on his lovely boat.

"Yes," he said, "It's pretty, but I don't like it. It's too tippy. If I don't stand in the center, it leans over. I saw you walking around in your boat, and it didn't tip much. I should have bought one like that."

Somehow or other, I had the brilliant idea that we might work out a trade. He was willing—even eager—and for the modest sum of about $50 I became the proud owner of the Old Town skiff that our family enjoyed for the next forty-nine years!

—Don Jordan, Glastonbury CT

Robertson's Riverside Boat House, at Riverside Station.

Robertson's Boat House at Riverside. The structure was torn from its foundations and damaged beyond repair during a tremendous flood in March of 1936.
Gray Collections

Robertson of Auburndale, Massachusetts. Stock was valued at ten thousand dollars and approximately fifty-two shares at one hundred dollars each were sold. The bulk of the shares was held by the Grays, the rest between Richardson and Robertson.

Whatever clout Robertson had, he must have used it greatly. When the deal was finalized, the corporation was now called the ROBERTSON & OLD TOWN CANOE COMPANY. The earliest catalog owned by the company dates back to 1902. It reads:

> Robertson & Old Town Canoe Company is a corporation, located at Old Town, Maine, and formed to combine the canvas canoe and boat manufacturing business of J. R. Robertson, of Auburndale,

Mass., and that of the Indian Old Town Canoe Company of Old Town.

> The new company is admirably located for this business in a section where canoes have been made and used by the Indians for centuries. It occupies a large, modern factory building fitted with improved machinery. Eastern Cedar, the best material in the world for canoe work, grows in abundance in the woods of Maine. Thousands of cedar logs come down the Penobscot River to Old Town; from these we select the best for our work.

The catalog cover lists "Canoes, Boats, and Yacht Tenders," and the interior of the catalog touted his twenty years' experience and the respect he had won from his many followers on the Charles River. The premier boat in the 1902 catalog was his design, the Robertson model, offered in five lengths, 15 to 19 feet. This canoe featured graceful lines and long decks similar to canoes popular in the Boston area. Prices varied from $36 to $42 in the AA grade. Most existing models took a $3 price hike over the previous year. The HW was offered as it had been in the past, but two models of the IF had been deleted, as well as the FB and GG models. The BB grade was also dropped, perhaps as an effort to eliminate customer confusion.

In addition to the canoes, the company also offered a canvas-covered dinghy or yacht tender designed by B. B. Crowninshield, a well-known naval architect from Boston. It was available in $11\frac{1}{2}$- and $12\frac{1}{2}$-foot lengths and included a rudder and bilge keels, one pair of rowlocks, and a pair of extra sockets, for the price of $46 and $50 respectively.

The *Old Town Enterprise* wrote little about John R. Robertson at this time, and it is difficult to determine where his allegiance lay. His business in Auburndale was well established before he signed on with the Grays, and during his association with OLD TOWN he kept his own enterprises operating. The *City Directory of Boston* lists a Robertson Canoe Factory at 132 Charles Street in the 1903 edition.

The *Enterprise* did report that the ROBERTSON & OLD TOWN CANOE COMPANY was running overtime trying to keep up with orders but was obliged to turn many away. "The flourishing industry has grown rapidly...," the paper reported on May 23, 1903. During that same eventful year, Sam, the only son of George Gray, graduated from Bowdoin College, where he had been secretary and treasurer of the Junior Assembly Committee, business manager of *The Quill* (the college paper), assistant manager of the Glee Club, and a member of Delta Kappa Epsilon. Besides his success in school, he was well liked in Old Town and was anxious to begin his career in the wholesale hardware industry, hoping to eventually work in Boston. His father had different plans. There was plenty of work to be done in the family operations in Maine, and Sam was expected to pitch in.

Whether it was because Sam was now available, or whether the Robertson affiliation hadn't worked out is not known, but it seems that almost as quickly as Robertson arrived on the scene, he hastily departed. Sam's hopes of working in Boston were abandoned; he was sent first to the woods, where he could learn the logging business from the ground up. He had spent many vacations working alongside his father at the

Bowman and Gray lumber camps, and now it was time for him to take a more active role.

He attended to the needs of the men, which was a challenging assignment, but it was just a small part of Sam's job. Equipment had to be maintained and sometimes replaced, horses purchased, and the woods cruised for a good timber stand. Sam went about learning all the aspects of the operation.

In August 1903 new incorporation papers were filed renaming the company the OLD TOWN CANOE COMPANY, with its officers listed as Herbert Gray, president; George A. Gray; and George H. Richardson, treasurer-clerk. There is no mention of John Ralph Robertson, who had returned to his ongoing endeavors on the banks of the Charles. For a brief period, ads encouraged would-be buyers to visit Auburndale, or to order a Robertson canoe from OLD TOWN, but this may have reflected a deal Robertson

Sam Gray, the son of George Gray, eventually assumed leadership of the canoe factory. Courtesy of Ruth Gray

This 1902 catalog reflects the new name but only appeared for one year. Gray Collections

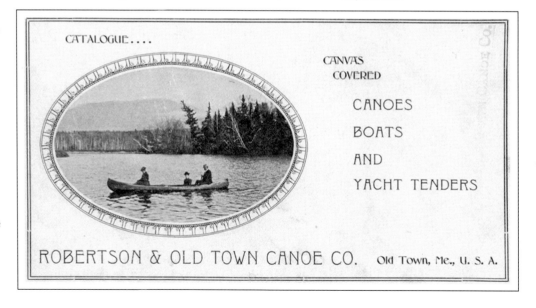

constructed upon leaving. Robertson continued to enlarge his operations at Riverside, and his penchant for speed led to sleek, new designs. But he did not limit his talents to the canoe business. His originality of design was evident in other projects. In 1905 he became involved in the production of the Stanley Steamer race car known as "The Teakettle," fittingly named for its two-cylinder, horizontal steam engine and 30-inch boiler. Robertson had been chosen by the Stanley brothers of nearby Newton as the builder of the car after they conducted their own drag tests. "They evaluated several models of canoes by towing each through the streets of Newton on a trailer hitched to a spring scale" to see which of the canoes produced the least drag. Robertson's canoes won. The car was built at Robertson's factory out of cedar strips covered with canvas and, on close examination, resembled a canoe turned upside down. The car's aerodynamics were further tested on the roof of the factory by running the long span.[2]

Robertson also supported racing endeavors in the American Canoe Association and became commodore of that prestigious group in 1908. After many years of successful boatbuilding on the Charles, he died on July 15, 1935, of a coronary thrombosis, and his ashes were sprinkled on the Charles in the area of Riverside that he loved so much.

If Robertson's departure from OLD TOWN in 1903 had an adverse effect on the company, it was never evident. The Robertson model canoe was renamed the Charles River model, and affairs in Maine seemed to settle back for another busy year.

Other canoe builders in Old Town were also making changes. E. M. White had moved to Water Street near the railroad spur in the heart of town in 1896. Carleton enlarged his factory at the end of North Fourth Street, preparing for a considerable increase in production each winter. Ingalls, who had been in business since the 1890s, sold his bateau interest to Carleton and planned to concentrate his efforts on canvas-covered canoes. His livery service was located right next to the Indian Island ferry, where a bateau made crossings to the Penobscot Indian reservation.

By October 1903 the OLD TOWN CANOE COMPANY was erecting a storehouse just above its factory, closer to the Bangor and Aroostook Railroad, to make goods more accessible for receiving and shipping. The railroad connections, so readily available in town, were a major factor in the company's success. OLD TOWN's advertising and marketing had a longer reach than any other Maine canoe company, and the railroad delivered its canoes to an expanding market.

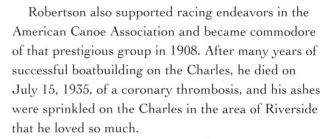

One of the Stanley brothers poses with Fred Marriott and "The Teakettle" after Marriott's unprecedented win in Daytona, where he set a world record for the mile at 28.2 seconds going 127.66 MPH.

Daytona Racing Archives

Above: THE ROBERTSON AND OLD TOWN CANOE COMPANY *staff gathers on the factory stairs in 1902. Some of the identities are no longer known, but the key figures are recognizable. Back row, upper left, John Ralph Robertson in his distinctive bowler hat; in front of him is George Gray. Far right, upper corner, Herbert Gray. Front row, center, Alfred Wickett with his younger brother Richard to his right. The second man up the stairs on the left is George Richardson, treasurer-clerk of the company.*
Courtesy of Braley Gray

Above: Although catalog drawings depict the ROBERTSON AND OLD TOWN *name painted on the factory, comparisons of photographs lead us to believe that the Robertson name never actually appeared on the building.*
Gray Collections

Far left: Old Town's ferry, a bateau, provided the public crossing to Indian Island. In winter when the ice was frozen, sawdust was applied across the ice to make the footing less treacherous. A bridge was finally built in 1951.*
Arthur Preble, Courtesy of Fogler Library, Special Collections

Dave Baker's restored 1916 Charles River model, Chameleon, manifests the beauty of the classic model.
David Baker

OLD TOWN's pride in its Indian heritage is visible on many catalog covers. On this page, covers from 1913 and 1914 (far right); on the facing page, from top left, catalog covers from 1915 and 1916; from bottom left, 1917 and 1918. Old Town Canoe Company

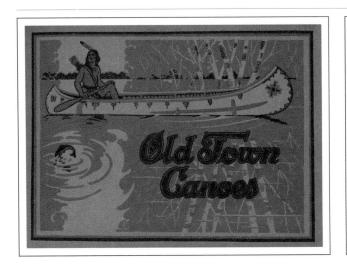

The OLD TOWN CANOE COMPANY catalogs were always important marketing tools. Some of the early catalogs were small, measuring approximately 3 by 6 inches. Although one of the first carried pictures of Robertson and touted his expertise, the introductions always contained a reference to the native Penobscots. The Grays relied on the Indians for part of their workforce, and in print, stressed this link to the Indians as a marketing strategy that gave the craft credibility:

As this innovation [canvas covering] had its origin here on the Penobscot River, it comes as a natural and appropriate sequence that the invention should be personified in the "Old Town Canoe." And the propinquity of the Penobscot Indians—but a river's breadth away, who are equally as well known as "Old Town Indians"—makes the name we have selected particularly apt. By our Indian workmen, and a number are included in our employees, there is infused into our canoe all that is possible of the old Indian romance, such as Hiawatha felt in the lines given us by Longfellow,

I a light canoe will build me,
That will float upon the water,
Like a yellow leaf in Autumn,
Like a yellow water lily..."[3]

Despite the diminutive size of the early catalogs, they offered much to interested buyers. Between their covers was a wealth of information: a foreword about the area and its tradition of building canoes, information on construction methods, and a plethora of options for the time. Canoes and boats were constructed of cedar ribs $5/16$-inch thick, 2 inches wide. The planking was $5/32$-inch thick by $3^1/2$ inches wide. Seat frames, thwarts, and decks were of mahogany, oak, ash, or maple, depending on the grade. Inner and outer gunwales were of spruce, unless ordered otherwise. Hulls were oiled before canvas was applied, adding resiliency. Different models required different weights of canvas. After the lead-based filler applied to the canvas had cured and was sanded again, the finished hull was painted and then varnished for extra protection and gloss. Early advertising emphasized the technical nature of the product, stressing that the canoes "were light on the portage...handled well in calm, rough, or white water...were dry and seaworthy...." Each model was fully described so that the discriminating buyer had an idea as to which canoe might be best for his purpose. (In some letters of correspondence, however, buyers

As the 1903 catalog emerged, the cover reflected a name change that would later identify the giant in the paddlesport world.
Susan Audette

OLD TOWN CANOE CO.

Catalogue
Canvas
Covered
Canoes

Boats
and
Yacht
Tenders

OLD TOWN, MAINE, - - U. S. A.

were encouraged to buy what was in stock.) It was company policy to follow up on every order to ensure customer satisfaction. Letters of testimony were used to tout the product, and they were very convincing. This was thought to be such an important advertising technique that the 1903 catalog had a separate insert devoted to letters of praise.

Harlem River, N.Y., June 8, 1901

Gentlemen:

Received the canoe in excellent shape and am very much pleased with same. Will be glad to recommend your goods to all inquiring friends and hope I may be the means of bringing you more business. Thanking you for favors received, I am,

Yours respectfully,

R. W. Weed

Indianapolis, Indiana, July 19, 1901

Gentlemen:

My canoe arrived yesterday in first class condition. Am well pleased with it in every respect. It is a beauty, rides the water to perfection, and has been greatly admired by all who have seen it. I think it is the prettiest boat around here, and there are over 100, and I am far from being alone in this opinion. Several have already asked me where I got it and also asked me for my copy of your catalogue.

Yours very truly,

Chas. S. Tilton

In OLD TOWN's 1905 catalog introduction, an increased sense of pride and commitment to quality was apparent in the text:

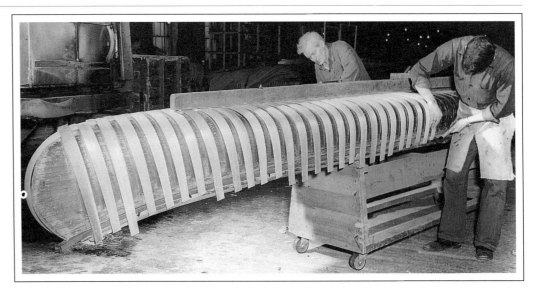

Workmen adeptly bend ribs onto the canoe form. Old Town Canoe Company

Ralph Nicola, Sr., tacks the stretched canvas to the hull. Jenny LaFrance

Our general plan of construction is the result of special thought, careful attention to detail, and painstaking plans for getting good results.

We will not put out of our factory a canoe to which we are ashamed to attach our name plate. This we mean shall stand for thorough workmanship, durability of construction, and all around good handling qualities. It is a matter of pride to us that one of our canoes shipped into new territory will bring us more orders.

While the idea, design and general plan of construction of the canoe was borrowed from the Indians, a colony of whom have lived at Old Town

War Canoes

The company introduced a smaller version of its war canoe in the 1909 catalog. A 25-footer, it was 44 inches wide, 14 inches deep, and sold for $75. War canoes were often a favorite in camps throughout the United States. You needed a crowd to get a big canoe moving, but it usually proved to be a lot of fun. It is thought that the term, "War Canoe" is a holdover from more turbulent times, perhaps the Indian Wars. They were briefly called Peace Canoes in the late 1960s, reflecting the tenor of the times, but they were never renamed in OLD TOWN's catalog. Whatever they are called, there is no mistaking one.

War canoes are still popular. In 1956 OLD TOWN constructed two 30-foot canoes for California's famed Disneyland.[4] These have been put to use as part of the Huck Finn Adventure theme park, where they are still used today. In 1971 a war canoe order was completed for King's Island Adventure Park, east of Cincinnati, which opened in 1972. The 34-foot canoes, made of wood covered in fiberglass, took 318 hours to construct.

Not to be outdone by previous generations, current workers at the factory began a group project in the winter of 1996–1997 to construct two war canoes to be used at the local Kenduskeag Stream Race, held yearly in April. One boat was constructed using the traditional materials of a wood-and-canvas canoe; the other was covered with fiberglass. Each canoe carried seven paddlers, including the CEO of Johnson Worldwide Associates, Ron Whittaker. This was Whittaker's first canoe race. The river was high, and although the paddlers of each canoe chose a good line, both canoes suffered the same fate at Six Mile Falls. Waves cascaded over the bows, swamping the canoes. Much to their shock, all the paddlers landed in the 33-degree water but were quickly rescued by support crews. Although a few suffered hypothermia from the cold conditions, most were none the worse for wear. The canoes, however, were not as lucky. Damage was extensive. Each canoe broke planking, seats, and thwarts, and 23 to 30 ribs. Since the incident, one of the canoes has been repaired, ready for a return run, but several team members seem reluctant to sign on again.

MICROBE CLUB Canoeing, Old Town, Me.

Above left: The launching of OLD TOWN's first war canoe was captured by the local papers and later distributed as a post card for the world to see. Gray Collections

Above right: The Luther Gulick camps on Maine's Sebago Lake supplied numerous opportunities for war canoe pictures. Gray Collections

Left: The 1920 catalog proclaimed, "Nothing is better in training for team work or unity of action and nothing is prettier than the even rhythmic sweep of a dozen or more paddlers driving a War Canoe with lusty strokes at top speed." Old Town Canoe Company

for centuries, modern methods and superior workmanship have brought steady improvement, until now, the canvas-covered canoe of today is a thing of beauty.

1905 also brought the introduction of a new model, for the moment simply named "Light Canoe." In response to repeated requests, the company designed this 15-foot canoe from the HW model. The Light Canoe was offered in CS grade and was only available in the "dead grass" color. Its Maine cedar planking was thinned to $1/8$-inch thick, and its ribs tapered at the ends and were spaced further apart. It had two seats, with one thwart for ease of carrying. Its price was $30, or $31 if a keel was desired. The canoe was designed primarily for long carries, where light weight was essential, and the catalog cautioned the reader:

> It is all right with proper care, but will not stand rough usage, banging about. Don't order it for a hard trip unless you know how and will handle and care for it properly.
>
> We have made some up ahead, but from the way orders are being booked the supply will not last long. Order early, if you want one.

By 1905 sales had already reached $25,000 and a new marketing strategy was tried to increase business.[5] Well aware of the growing interest in recreational canoeing to its south, in 1906 OLD TOWN CANOE began to advertise to buyers who would use the canoe for fun. In one-inch advertisements in weeklies with national circulation, OLD TOWN's ad copy stressed the new angle: "the pleasure to be derived from owning a canoe." The response amazed the Grays, and they knew they were onto something—a whole new group of buyers.

The 1906 catalog introduced OLD TOWN's first "war canoe," a large canoe that held several paddlers. The first of its kind was unveiled on April 21, 1905. The 35-footer had made its maiden run on a clear Sunday afternoon at the riverfront not far from the factory. Twelve men in suitcoats and bowler hats, seated on 4-inch thwarts spaced about 27 inches apart, tested the canoe's seaworthiness and as the *Bangor News* reported, "an admiring crowd of onlookers inspected the craft and the general opinion expressed was that it was a 'beauty,'...practically perfect." The canoe was to be used at a summer resort called Broad Ripple and was shipped to G. E. Graham, in care of the Microbe Club in Indianapolis. Although the original boat was reported at 35 feet, a 34-footer was described in the 1906 catalog. It offered one stern seat, and the planking and ribs were extra thick. There were floor braces and a keel. The war canoe's absence from earlier catalogs may have been due to lack of space.

It seemed that each year the company offered something new. One modification, in particular, was revolutionary: open gunwale construction. This was first seen on the 1907 Ideal Canoe, available in the AA grade of the 16-foot and 17-foot Charles River models. Prior to its introduction, canoes were all finished with closed gunwales, and when the boat was stored upside down, water did not drain easily, causing the gunwales to rot. The new method finished the canoe with inner and outer gunwales, which were rabbeted to cover the

It was a hot Saturday afternoon in July as I stood with a friend under the giant white oak that welcomes visitors to my home. Conversation was about canoeing and a fall trip to Northern Maine, when slowly a vehicle approached with a tattered canoe atop the roof. Another canoe had found my small shop, but my first assessment indicated it would not be staying. While an elderly gentleman introduced himself to me, I gazed through two large holes in the canoe, saw its fiberglass covering, broken mahogany gunwales, thwarts and seats, rotted stems and decks, and what would require the replacement of 25 ribs and 75 feet of planking. The canoe had had a serviceable life, but it had moved into the past tense.

The man pleaded with me to restore the canoe, even knowing the expense involved. The canoe was a 1925 OLD TOWN AA Ideal, a boyhood gift from his father. It had been used at the family retreat, and he had passed it on to his own son, who had passed away prematurely. The canoe had fallen into disrepair.

I reluctantly took possession of the canoe, and it lay idle for several months, devoid of soul or character. Work began in earnest in December, and slowly its identity began to reappear. It developed character, and by the end of summer its soul had emerged again.

As fall approached, I stood under the canopy of green leaves again, soon to surrender to autumn's pressure, and awaited the return of Mr. Walker. There was no need for words. The canoe I had restored spoke for itself. Mr. Walker's eyes revealed a dichotomy of emotions, from happiness to sadness. The canoe contained the memories of a lifetime and future memories not yet recorded. As he drove away, I felt an emptiness that I could not fully share Mr. Walker's emotions, but smiled, turned, and walked away, knowing part of my soul was with his Old Town canoe and the memories he deeply cherished.

—Ken Morrisroe, Brooklyn, CT

A beautiful day, an attractive couple, and an OLD TOWN OTCA.
Old Town Canoe Company

top edges of the planking and canvas, leaving the tops of the ribs exposed, thereby creating drainage holes between them. This extended the life expectancy of the gunwales considerably.

The Ideal's second new feature was half ribs between the full ribs. This construction technique became an option on other models and eliminated the need for floorboards to protect the planking. (E. M. White had utilized this technique in the early 1890s.)

Competition was fierce. New models and innovations helped keep the public interested. Society was on the move, with more leisure time and a yearning for speed. OLD TOWN had exhibited a gasoline-powered canoe as early as 1905, at the New York Sportsmen's show. Its engine was manufactured in Old Town at the T. M. Chapman and Sons foundry, a company that began its business in 1851 making sawmill machinery and later expanded its products. The 1908 catalog introduced the Power Canoe Boat,

*OLD TOWN CANOE combined efforts
with T. M. Chapman and Sons, engine
manufacturers, to produce the Power
Canoe Boat in 1906.*

Gray Collections

which was a remake of the XX Double End Boat with a slightly different twist. This 18-foot craft was powered by a $^3/_4$- or $^1/_2$-hp Chapman gasoline engine, but the company was willing to install any engine the customer preferred. The boat could also be equipped with four folding cane chairs which could be positioned to adjust the boat's trim.

Another new offering in 1908 was OLD TOWN's OTCA model (an acronym for OLD TOWN CANOE). This design integrated the best features of some of OLD TOWN's proven models and had open spruce gunwales and graceful, 20-inch-long decks with a low coaming. The bows were full like the HW model's, the floor flat like the Charles River's, but the beam had been increased slightly to provide more stability without a proportional loss of speed. It was offered in 16- and 17-foot lengths in stock green, and although it

My wife Jeanne always said she married me for my sailing canoe, an 18-foot 1919 OLD TOWN OTCA. Oh, incidentally, we visited the OLD TOWN factory on our honeymoon and, yes, we're still married —R. Rybinski, Manlius, NY

As I was putting the final coat of green paint on our restored OTCA, her beauty struck me. She was gleaming in that southern sun, just hankerin' to be put in the water. Saturday was the trial run on the canal in downtown Augusta, Georgia, for the annual Public Radio Canoeathon. To alleviate parking congestion, the committee asked if we would leave our canoe off Saturday evening. I said, "What? Not a chance!" Even with a guard, there's no way I'd do that. We arrived very early Sunday morning. —Julie McCrum, Aiken, SC

Silence is one of the most endearing traits of a wood and canvas canoe. Nothing glides so quietly through the water—but it can also create quite a commotion! My restored 1948 OTCA frequently draws admirers and attracted one of its largest crowds in 1990, when my wife and I paddled the Allagash River with our two young sons. At the head of a five-mile rapid called Chase Carry, our canoe was completely surrounded by people, including several old-timers who reminisced about the days when wood and canvas canoes were the only ones on the river. Younger paddlers couldn't believe I was going to take the OTCA through the rapids with my ten-year-old son in the bow. They thought the canoe should be kept at home in a place of honor, where it wouldn't get ruined. We discussed the great qualities of strength and flexibility in wood and canvas canoes, and then departed down the rapids. We did hit a couple of rocks, but the canoe wasn't damaged in any way. The next day we encountered a Kevlar canoe with its snout wrapped in duct tape. It, too, had hit a rock in Chase Carry, and it cracked like an egg! —Lyn Bixby, Storrs, CT

My mom and I used to take our 18-foot OLD TOWN OTCA and go skinny dippin' after dark. To get back in the canoe, I could slip right over the side just like a wet seal. —Mrs. Leonce Rousselot (88 years old), Cooperstown, NY

was the CS grade, it was thought to be as attractive as the Ideal model—but at a lesser price. Editing notes that appeared in the following year's catalog state, "This model was introduced last year and it proved so popular that we had difficulty in filling orders.... This canoe has the most steadiness of any model we offer."

With the continuous introduction of new models and innovations, supported by extended advertising, sales at OLD TOWN CANOE were up 25 to 30 percent over the previous year's figures. The inside cover of the 1908 catalog identifies the extent of OLD TOWN's business by impressively listing its foreign agencies in "Buenos Ayres [sic], South America; Paris, France; Hamburg, Germany; Abo, Finland; and Firenze, Italy. The Pacific Coast was equally well covered with agents in San Francisco, Seattle, and Portland. OLD TOWN was becoming a global name.

1   ROBERTSON & OLD TOWN CANOE COMPANY Catalog, circa 1902.
2   Dick Punnett, *Racing on the Rim* (Ormond Beach, Florida: Tomoka Press, 1997), p. 62.
3   OLD TOWN CANOE COMPANY Catalog, 1906.
4   *Bangor Daily News*, April 24, 1956, p. 6.
5   Gray, "How We Built New Markets," *System*, p. 58.

7

Labor Pains

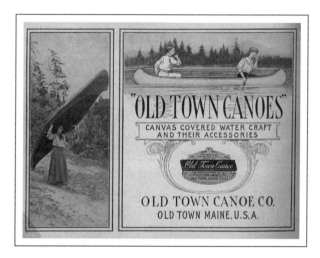

With the increasing popularity of the canoe in the early 1900s, competition in the industry was keen. Although there were several builders in town besides OLD TOWN CANOE, only the E. M. White Company was in close competition. Rumors of any dislike or distrust between the two companies were unfounded. They were certainly very aware of what the other was doing, but White did not attempt to compete on the same scale with the growing OLD TOWN. He was a builder, and the Grays and their partners were businessmen. The two companies were cordial to each other, sharing materials now and then, but still keeping a watchful eye on each other's new offerings at trade show time.

"Canoes by the hundreds" was certainly the reality in Old Town and the prophecy of the *Old Town Enterprise*'s editor had greatly underestimated the canoe craze. Canoe builders were working overtime, and with all the increased demand for skilled workmen, the labor force began to feel the pinch.

The biggest threat to OLD TOWN CANOE came from the Chestnut Canoe Company of Fredericton, New Brunswick. Despite little previous contact with each other, the companies found themselves in court on opposite sides. On December 21, 1905, a lawsuit initiated by the OLD TOWN CANOE COMPANY accused Chestnut of stealing employees. It asserted that Allie Ayers and nine other OLD TOWN CANOE COMPANY workers had been enticed by Chestnut to leave OLD TOWN and go to work for Chestnut in Canada. OLD TOWN claimed this to be a violation of the Alien Labor Law, which prohibited citizens from working in another country.[1] This was the first case of its kind in Canada, and Chestnut was liable for a maximum penalty of $1,000.[2] But after the case was heard, the minimum fine was levied and Chestnut was only required to pay a total of $50.

The case was overturned in January 1906 when a higher court heard an appeal questioning the judge's jurisdiction in hearing the case. Lawyers' fees ended up costing OLD TOWN in excess of $225, much more than the original decision.[3] Four of the men stayed on in the employ of Chestnut, including Richard Wickett, in spite of the fact that his brother Alfred was acting superintendent of the OLD TOWN CANOE COMPANY. OLD TOWN had been attempting to protect its labor force with the lawsuit, and in spite of its losses, had made a point.

Collectively, the builders in Old Town were thought to be manufacturing 2,500 canoes over the 1905 season.[4] Shipping that quantity would require eighty railroad cars with thirty canoes in a carload. OLD TOWN's ledger indicates an increase in shipping costs. Each maker in town seemed to have his own special niche. In addition to canoes, Carleton had an order from Commodore Robert E. Peary for three steel armored bateaux, which were to be used on an Arctic region trip, and White was building motorboats.

Production continued to increase at OLD TOWN CANOE, and George Gray's logging operation attempted to meet the demand of the growing company. During one of the drives in 1905, he and ten men were working four million feet of logs downriver, to be used for "siding, shingles, railroad ties, and building materials, as well as wood for the canoe company."[5]

The greatest impact of this increase in OLD TOWN's

E. M. White offered a full line of canoes and boats. Some are shown here at Indian Landing in 1906.
City of Old Town, 1906 Souvenir Book, *Courtesy of Fogler Library, Special Collections*

business was the need for space. Bickmore's Gall Cure, whose own production continued to grow, left the fourth floor of the factory to open its own facility on South Main Street. The building still stands today, where it houses several apartments. Its old logo, a horse named Doctor, can still be seen above the door. Doctor was a heavy draft horse owned and worked by the Penobscot Chemical Fibre Company. A dapple gray with good height, a deep chest, and powerful legs and back, he was good-natured and willing. Unable to work due to open wounds, he was treated with the Bickmore salve successfully. From that day on, he represented the product as its goodwill ambassador whenever possible.

Reorganization at OLD TOWN CANOE was not limited to adding space, and new incorporation papers were filed in 1907 listing Herbert Gray as president; George Gray; George Richardson, clerk; and Sam Gray, treasurer. Alfred Wickett was named superintendent of the company.

The 1908 catalog boasted that they were even "bet-

The Chestnut Canoe Company began with Jack Moore and Allan Meads building custom canoes and boats for the R. Chestnut & Sons hardware store in the late 1890s. While many canoe companies of Canada served regional needs, Chestnut reached out to international markets as well.

Rodger MacGregor

ter prepared than ever before to give efficient service." Attention to detail and to the needs of the customer were continuous goals, and models continued to be redesigned to meet the whims of the public. OLD TOWN had nearly doubled its factory space and had a railroad spur running directly into the plant so it could load cars directly. According to the local paper, thirty-five men were employed in the plant.

By 1910 its workforce grew to sixty individuals employed year round, and an earlier building expansion program was well underway. The Grays had decided to add a large brick building approximately 150 by 50 feet at a cost of $6,000.[6] Shipping was now completed on the basement level with an easy access

to the driveway. The top two floors were used for storage of canoes. Included in the project was an additional large wooden storehouse. During construction of the new brick building, George requested two windows be added to allow more light. The contractor was willing to comply, but the additional cost seemed exorbitant to George and infuriated him. From that day on, George Gray resolved that any new construction would be done by his own men.[7]

Sales orders taken throughout the winter indicated that a great season was about to occur. By February of 1910, 2,000 canoes were in stock to meet the upcoming demand. All the area companies

---

Bickmore, Still Thriving Today

Bickmore's Gall Cure—the miracle cure that helped heal sores on horses and oxen caused by the abrasion of harnesses, yokes, saddles, and other equipment—had a meager beginning in 1884 when Abiel Parker Bickmore formulated his salve on the stove in his kitchen. Abiel Bickmore had no idea that the product and company would still be thriving today.

As use of the horse waned, Bickmore attempted to diversify by adding other health and beauty products: XYZ ointment for sunburn and insect bites; Neatslene Shoe Grease for waterproofing; Gold Star Insect Repellent; shaving cream; and a mortician's powder.

Although the company held on, profits were low. It was finally sold in 1969 or 1970 to Abe Siegel of Skokie, Illinois, who bought the firm for his son when the company was grossing only $30,000 a year.

Siegel's son was apparently uninterested in his father's acquisition, and the company was sold to Peter Ottowitz, the present owner, who not only has brought the Gall Cure back to its previous prominence, but has added new products as well. They include Bick-4 Leather Conditioner, Bick-1 Leather Cleaner, Suede & Nubuck Cleaner, Gard-More Water and Stain Repellent, Ultra-X Hat Cleaner, Kahl & Sons Felt Hat Stiffener, and Bickmore's Creme Polishes. Sales of the Gall Salve continue to grow—mostly by word of mouth. The balm remains one of its "showcase" products and the leader in its animal health care line. The Hudson, Massachusetts, company now grosses over one million dollars a year, with marketing and sales once again covering the globe.

seemed to be sharing the prosperity, but none as greatly as OLD TOWN. The Labor Statistics of Maine tell the story: in 1910 Carleton employed 20 men, OLD TOWN reported 50, White 9, and Morris 30 men and 1 woman.

By 1910 OLD TOWN had reintroduced the Light Canoe, now renamed the Fifty Pound Model. Its reappearance reflected the public's demand for a lightweight, utilitarian canoe. The 1910 catalog explained that the Fifty Pound Model was ideal for one person with a heavy load or for two men with a light load. Ribs were the standard size but tapered at the ends and were spaced $1^3/4$ inches apart; planking was thinned to $1/8$ inch. The canoe was now equipped with a removable middle thwart for portaging. Canvas was number 10, quite light compared to other models, and the canoe was painted without the topcoat of varnish. Although this was a canoe the public wanted and the company had been making for the past five years, it upset George Gray and his men to make it. George felt it was a waste of good material to plane the planks thinner. His men found the construction tedious, and the extra labor involved lowered their piecework production.

Always with an eye to the future, Sam Gray acquired the Carleton Canoe Company in 1910. Carleton's sawmill had supplied much of the cedar to OLD TOWN's canoe shop, and the addition of another sawmill was a strong business move. The Carleton plant was located adjacent to the boom and the necessary supply of wood. Carleton's office and display rooms in the Ounegan block by the falls, and a shop on lower Main Street, were all included in the deal. It was

Doc, a gray Percheron, became the logo of the Bickmore company and is shown here over the door of their new factory, built in 1906.
Susan Audette

A Bickmore dealer lets the advertising do the talking for him.
Peter Ottowitz

Beautifully balanced

"OLD TOWN CANOES" are the exact reproductions of models built by the Penobscot Indians. These Indians were masters in the art of canoe building. Their canoes were speedy, they carried large loads easily and—they were beautifully balanced.

"Old Town Canoes" are not only steady, fast and beautiful—they are also light in weight and remarkably durable. Sturdy and rigid, "Old Town Canoes" stand up under the severest strains—they last for years without repairs.

"Old Town Canoes" are low in price too. $64 up. From dealer or factory.

The 1926 catalog is beautifully illustrated with all models in full colors. It gives prices and complete information. Write for your free copy today. OLD TOWN CANOE COMPANY, 694 Main Street, Old Town, Maine.

"Old Town Canoes"

Nature Magazine, *April 1926.*
Sue Audette

OLD TOWN often employed Penobscot Indians to attend some of the larger trade shows in New York and Philadelphia. The group would perform Indian songs, tribal dances, and tell Indian myths. On display as well was the "distinctive Indian craftsmanship which has ever been present in Old Town Canoes and made them the outstanding canoe the world over." Seen here are (counterclockwise) Roland Nelson (upper left), John Susep, known as Basshorn and whose face became the icon used in many catalogs, Lena Polcies, Louie Nichola, and an unidentified woman. Gray Collections

Sam's intention that the Carleton business continue under the same name using the same production methods. By acquiring the Carleton Canoe Company, the Grays were able to offer more products with the know-how and equipment already in place. An announcement issued from Carleton's new management read:

On March 22nd a new ownership took possession of the stock, good will, all rights and property of the Carleton Canoe Co. The former owners desired to retire from active business and the new control, continuing under same name as before, aims to enlarge the plant, improve the products, and increase the sale of "Carleton Canoes" and equipment.

We shall seek to merit a continuance of your

patronage and in a few days the delay incident to the operation of our shops and the attention to correspondence due to stock taking and readjustment to new managership, will be overcome and immediate action given to your inquiries and orders.

New Catalogs will be ready to issue inside of ten days when we will send you one.

May we have the assurance of your continued support for "Carleton Canoes"?

—Old Town, Maine, Carleton Canoe Company
March 25, 1910

With the factory expansion nearly complete, the addition of Carleton, and improvements being made to the existing plant, Sam Gray left for a sales trip out West in November 1910. Each winter's sales trip reaped numerous orders for the coming year. Sam returned home pleased that his efforts were paying off.

By 1911 the catalogs expanded to a 6- by 8-inch size. The catalog and Sam's ad campaigns reflected Sam's well-thought-out strategies. Glowing testimonials covered a page or more, and the catalog warned buyers to look for the OLD TOWN trademark and not to accept any substitutes. Sam's advertising genius was seen in his neverending attempts to bring in new buyers. There were enough OLD TOWN canoes out in the marketplace in all kinds of service so that the technical aspects no longer had to be stressed in the ads. Instead, Sam turned his attention to targeting special markets. By being more sensitive to societal changes, he developed ads to address segments of the population, attracting special interest groups with a personalized touch.

For instance, in women's magazines, the ads

We got the 15-foot OLD TOWN Light Weight canoe from a barn, where it was being used as a clean-up receptacle for horses and cows. After shoveling it out, squirting it with a hose, and cleaning it (doing wonders for the flowers at the foot of the driveway), we restored and re-canvassed it. We didn't have to put any linseed oil on the wood, as it seemed to have plenty of life. The inside now has a much deeper color than the normal cedar boat, but with a satin varnish looks wonderful! One of the members of our wooden canoe restoration group of The Chesapeake Wooden Boat Builders took it to the mid-Atlantic wooden boat show in St. Michaels, Maryland, and it won second place in the restoration category!

—A. Gillis, Glenelg, MD

I was visiting some old colleagues who wanted to get rid of an old canoe. They knew they had a "live one" when I accepted, sight unseen. When I did see the canoe, I could see right through it from fifty feet away! But it was an OLD TOWN, and I had agreed to cart this old pile of wood away.

I called Gil Cramer, a friend who restores canoes, and he agreed to look at it, if not burn it. Another friend agreed to drop it off at Gil's if I would cover any fines he got for littering the highway if the canvas began blowing off along the way. A couple of months passed before I heard from Gil: "Well, that little canoe is going to take thirty-two ribs, new decks, inner and outer rails, and a lot of planking, but she is going to be a real cutie!" A deal was struck, and he agreed to deliver the canoe at the WCHA meet in Paul Smiths, New York, that July.

I first saw her in a mist at dusk. I thought there was no way it could be her. Wow, was she beautiful! I couldn't wait until morning to see her in the daylight and paddle her. Gil also wanted to see what she would do. I was so excited I was giggling, and that little canoe took to the water like she was giggling, too. With the first stroke of the paddle, I knew I was in big trouble. I couldn't part with her—she was light, fun to paddle, and I could flip her over my head like the guys. She talked to me all the way across the lake and we fell in love. If she was going home with me, I had to figure out a way to get her past my husband, so I gave her to him (or so he thinks) as an anniversary present in August. It worked! She is now part of the family, with our eldest son now laying claim to her. This little OLD TOWN Fifty-Pounder lets us know she's delighted to be a "keeper" every time we take her for a spin!

—Julie McCrum, Aiken, SC

showed a lovely lady paddling in the bow. She was fashionably attired, emphasizing her femininity, yet because she was paddling, she paraded her independence. The ads emphasized the beauty and finish of the canoes, the ease with which they could be paddled, and stressed the steadiness offered in OLD TOWN's designs, reassuring mothers and wives alike. A woman portaging a canoe was even featured on the 1911 catalog cover.

He also sought prospective buyers with copy aimed at summer camps, Boy Scouts, fishermen, campers, and seaside resorters. He tempted the summer-resort-

Introduced in 1905 as the Light Canoe, and renamed the Fifty Pound Model in 1910, this canoe had a removable middle thwart for portaging. This Ozzie Sweet photo, used on the 1962 catalog to demonstrate the ease of portaging, was also used in a Canadian postcard series and by L. L. Bean for one of its catalog cover drawings.
Gray Collections

goers with a description of blue skies and the calm stream, an idyllic day on placid waters. Other ads extolled the camaraderie of paddling a war canoe with ten or more companions.

Most of all, Sam never forgot his roots. The catalog introduction always touted the company's association with the Indians, whether in a full-page dissertation or with a small picture inset of one of the local Penobscots. In *Boy's Life*, ad copy stressed that OLD TOWN canoes were patterned after the real Indian models.[8] Sam believed that all boys found some appeal in the word "Indians." He also used pictures of famous people and their testimonials to good effect.

The year 1911 was to be hot in more than just sales. Spring began with a severe drought, and fire warnings were issued on the front page of the *Enterprise*, harbingers of the huge fire in Bangor that followed. Old Town was not spared. On May 17, 1911, the Carleton Mill burned. It was nearly gone before the firemen could get there. It supposedly started from a "hot box," an overheated gear at the end of an axle. Fifteen men were working on shingles at the mill, and they managed to save the canoe and bateau stock in the adjoining building. The townspeople worried that the mill would not be rebuilt, but the civic-minded Grays were in Old Town to stay. The mill was rebuilt on property on the west side of the Penobscot River above the Conant and Carr Mill, and was soon up and running again. The OLD TOWN CANOE COMPANY now owned all the property on the east side of upper Fourth Street. On September 2, 1911, the *Old Town Enterprise* announced the Grays' latest plans:

Old Town Canoe Company will be moving a two-story wooden structure next to the Bangor and Aroostook track to the north 200 feet. In its place they will construct a four story brick building 200 feet long for canoe manufacture. Capacity will make production of 10,000 a year within distance. They will manufacture about 4,000 this year...and probably will employ 100–105 next season.

1 "Old Town Man Sues Fredericton Canoe Maker," *Bangor Daily News*, April 12, 1915, p. 3.
2 "A Novel Law Suit: The Old Town Canoe Co. Figures in Canadian Courts," *Old Town Enterprise*, April 15, 1905, np.
3 Documents courtesy of OLD TOWN CANOE COMPANY.
4 *Old Town Enterprise*, July 1, 1905.
5 *Ibid.*, April 22, 1905.
6 Jim Cunningham, Interview, former superintendent of the OLD TOWN CANOE COMPANY, May 1993.
7 *Ibid.*
8 Gray, "How We Built New Markets," *System*, p. 58.

Canoeists paddle on the river in front of the Carleton Mill before it burned and was rebuilt.

Gray Collections

George Gray had his own crews complete Building #3 in 1912. OLD TOWN CANOE COMPANY was certainly changing, but not only with its expansion. Alfred Wickett, who had been the first builder for the company, sold his stock in 1914 to the Gray family and left their employ. Soon, he was the glory of Milford, the town across the Penobscot River from Old Town. There, in 1915, he opened the Penobscot Canoe Company, manufacturing canoes, yacht tenders, and double-ended rowboats. The canoes looked much like OLD TOWN's, but Alfred instituted a design change that he felt added strength and rigidity to the end of the canoe. Instead of the inside gunwale being carried to the very tip, he invented a spearhead-shaped deck that he used on his Penobscot canoes. The innovation received a patent on August 22, 1916.

At that time, he was already employing about ten workers and had a hundred canoes in stock. In the spring of that same year, he incorporated his business with a new partner, Henry Barker, who owned a successful lumber company in town. According to the local press, the Penobscot Canoe Company was doing well, and there were hopes that it would be as prosperous as "its sister across the river."[1] In late 1916 a 25-foot canoe was being built for an order from South America, and by November the company moved into new quarters. Wickett, as manager and chief salesman, was making good use of his OLD TOWN experience to establish his new business.

The Penobscot Canoe Company seemed on its way, but corporate papers were not filed annually for the years 1917–1921 and seemed a precursor of trouble.

8

Alfred Wickett Departs

Wickett's marriage to Gertrude Thompson ended in divorce in April 1918, but by November of the following year he married Myrtle Bamford, a local woman who ran a nursing home in Old Town. The Barker Mill experienced a fire in April of 1920, and the canoe company sent a small crew to a local snowshoe plant—the Penobscot Snowshoe Company—where they were able to finish up canoe orders. But with the complications caused by the fire, the breakup of his second marriage, and the pressure of the business, Alfred abandoned his canoe business to his partner and left to make another start.

First he traveled to Fall River, Massachusetts, and during World War I found employment in the shipbuilding industries. From there he traveled westward, finally landing in Valley Park, Missouri, a resort-like town bordering the Meramec River.

The timetables are not clear, but it appears that Alfred was already well-established and living in Missouri when the Penobscot Canoe Company burnt to the ground on November 5, 1923. According to some, the blaze was suspicious, but no proof of this claim can be found. The insurance companies attempted to settle for $18,000, but this settlement was adjusted after a court fight and resulted in $24,500. Although Wickett's name was mentioned in newspaper reports of the proceedings, it was not clear if he still had interests in the company or received any part of the settlement.

Valley Park provided a new beginning for Wickett. Located just west of St. Louis, this young town was home to the St. Louis Plate Glass Company and many other lucrative businesses, and Alfred found a ready public. As early as 1902, canoeing had become a popular activity on the Meramec River. Daily, the Frisco or Missouri Pacific train lines carried passengers to Crescent, a town southwest of Valley Park. Once there, canoeists paddled downstream to Valley Park to board the trains home.

Canoe clubs were already established by the time Wickett arrived. The Paddle and Saddle Club had its headquarters in a commodious stone building on the banks of the river. Besides canoeing, members enjoyed games of polo, cross-country rides, and tournaments of all sorts. The Meramec Canoe Club, located near the city, offered much the same. The town boasted factories, low taxes, and reasonable fire insurance. It had an up-to-date orchestra, more than six hundred skilled workmen, and eight saloons of the quiet, law-abiding, orderly kind. It was truly the perfect place for Alfred.

A trademark building feature of a Penobscot canoe is the spearhead deck formed by tailoring the inner gunwale. On Wickett's later canoes, built by the St. Louis Meramec Canoe Company, the scarf joint reversed the design to an arrowhead design, resulting in a much stronger joint.

Gray Collections

THE PECACO IMPROVED DECK

PATENTED AUG. 22, 1916

Supplied in all models
See opposite page

Here he began to do what he did best and soon opened the St. Louis and Meramec Canoe Company, manufacturer of the Arrowhead Canoe. The designs were Alfred's but with one minor change in the deck to make it even stronger. Instead of the Penobscot spearhead, the Arrowhead Canoe would bear just that, a deck in the shape of an arrowhead with an improved scarf joint. Wickett's 1925 catalog boasted:

Canoeing is an original American sport and the first mode of water transportation in America.... "Arrowhead Canoes" are vastly superior in every way to the original "Indian-built canoes." They are safer, stronger, more graceful, and with proper care and use will last a lifetime.... All Arrowhead Canoes are designed and built under the personal supervision of Mr. A. E. Wickett of Old Town, Maine, who has had 35 years' canoe building experience.

Several models—including racing sculls—were added to the extensive line of canoes and canvas-covered boats. The cost of his basic Meramec model canoe in 1925 began at $68 for a 16-footer, to $78.50 if half ribs and outside stems were added. Options included striped edges, and a choice of green or red paint. Other colors required an increase of three to four days to complete the order. Sponsons were also available. Wickett had certainly found his niche.

In 1930, after several successful years manufacturing, Alfred married a third time to Sophie Hodnett, the owner of a local hotel in which Alfred had lived. His son by his first marriage, Laurence, joined him in Valley Park, where together they opened the St. Louis Boat and Canoe Company, expanding the canoe pro-

"23 DON'TS"

Now that you have been told "what to do," some advice as to "what not to do" is just as much, if not more important.

(1) DON'T take a canoe out alone without first learning how to handle it.
(2) DON'T take passengers out with you until you know how to manage your Canoe.
(3) DON'T think that a canoe can't be tipped over.
(4) DON'T overload a canoe.
(5) DON'T take unnecessary or hazardous chances.
(6) DON'T try to change seats or positions in a canoe (unless you are an expert).
(7) DON'T ever try to carry more than four persons in a 16-foot canoe and then only in calm water.
(8) DON'T try to carry more than two persons in rough water.
(9) DON'T grab hold of the gunwales if the canoe should rock or tip. Keep your weight always in the center and rigid, the canoe will come back.
(10) DON'T take persons out in a canoe that have a great fear of the water or of canoes. Show them "how safe" it is first and that you are "master" of it. This will overcome their fear and they will not be apt to get panicky if it rocks a little bit. Canoes cannot help but rock anymore than an automobile can go over a bump or bumps without rocking. Can you imagine an automobile going over a rough road as smoothly as a canoe does over rough water?
(11) DON'T ever get into a canoe facing any direction other than the way you are going to sit or kneel.
(12) DON'T ever sit in the stern seat if paddling alone, sit in the bow seat facing the stern or kneel just ahead of it, or just a little back of the middle.
(13) DON'T load a canoe so that the bow end rides lower in the water than the stern. The bow should be a trifle higher.
(14) DON'T sit on the seats when paddling in very rough water, get on your knees and "paddle like the Indians" did. You can steady the canoe and "ride the waves" better from this position.
(15) DON'T fail to keep a floor rack in your canoe and step or walk on it always.
(16) DON'T loan your canoe to anyone who does not know how to handle it. You wouldn't let someone take your auto that did not know how to drive or you wouldn't expect anyone to let you take theirs if you did not know how to drive.
(17) DON'T do anything that will jeopardize your own life or the life of another. Do your experimenting in a bathing suit, where and when it is safe.
(18) DON'T ever try to be "smart" by rocking the canoe to annoy or scare a passenger. They might do something that would surprise you and "canoeing will lose a good prospect."
(19) DON'T make promises to a passenger and then break them. Life long enemies of canoes or prejudices against canoes are made in this way.
(20) DON'T take passengers who cannot swim out in rough water, or don't unnecessarily get in the waves of big boats or launches.
(21) DON'T forget that canoes have never yet tipped over "when empty," proving that it is the people in them that do the tipping.
(22) DON'T fail to overhaul your canoe annually.
(23) DON'T think a canoe is made "fool proof."

After you become experienced in handling your Arrowhead canoe, you will be as much "at home" in your canoe and on the water as you now are on land. Any further inquiries with reference to management or handling of canoes will be answered by Vincent M. Smith. Enclose a stamped envelope with your inquiry.

It should be interesting to know that Vincent M. Smith has been paddling a canoe for a quarter of a century and has won every race in which he has entered in the last three years. He is the present holder of the Gardner Trophy for the half mile singles of the American Canoe Association, W. D., which he won in 1922-23-24. He is the first one to win this trophy three times. He is also holder of over one hundred trophies, cups and medals won in canoeing. Mr. Smith is manager of our sales office at 192 N. Clark St., Chicago, Ill.

ST. LOUIS MERAMEC CANOE CO.

Page Two   Valley Park, Mo.

Commonsense don'ts, seen in the 1925 St. Louis Meramec Canoe Company catalog. It always pays to remind the customer.
Gil Cramer

duction to include retail sales of other brands.

"Chummy," as Laurence was called, had an affinity for speed. He satisfied that interest by opening his own retail store in St. Louis, called the St. Louis Boat and Motor Company, just a few blocks from the Mississippi River. With his new venture, Chummy concentrated on motorized craft while his father managed his own canoe manufacturing business. Laurence's interest was fueled by the growing popularity of motorboats and the improvements that could be added to them; his 1955 obituary credits him with "designing and originating many marine items, including refrigeration and air-conditioning now commonly in use on large boats."[2] Both he and his father were well known in the St. Louis area, and respected by the boating world.

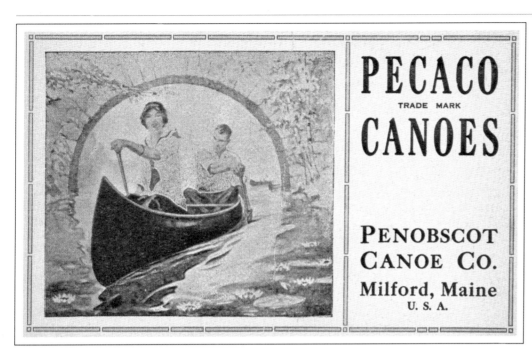

PECACO

TRADE MARK

CANOES

PENOBSCOT
CANOE CO.

Milford, Maine
U. S. A.

A PECACO catalog, circa 1920. The
local newspapers hoped that the
Penobscot Canoe Company would be
as successful as its sister across the
Penobscot River.
Gray Collections

After a long career, Alfred died on September 14, 1943, from complications incurred from a ruptured appendix. His obituary recognized him for his many achievements and credits him as the "designer of the first canvas-covered canoe." Although this isn't true, he had probably been instrumental in producing some of the earliest commercially produced canvas canoes, and his efforts helped to shape the industry. His gravestone in the Oak Hill Cemetery in Kirkwood, Missouri, reflects what he loved best and bears a deeply etched carving of a canoe.

[1] *Old Town Enterprise*, August 19, 1916.
[2] *St. Louis Globe-Democrat*, September 2, 1955.

W ickett's departure from the OLD TOWN CANOE COMPANY in 1914 had little impact on the firm. Production was running smoothly, and there were other capable workers who were tapped to fill his shoes. The greatest concern was filling the orders in the space available, in spite of the fact that two new brick buildings had been erected not long before. A boiler room with an 80-hp boiler—the most modern of its time—capable of burning scrap lumber had also been installed, along with electric lights and a $3,000 sprinkler system.

Though unrest in other parts of the world caused some concern, it was not immediate or near enough to dampen the enthusiasm in Old Town for further expansion. On July 18, 1914, the *Old Town Enterprise* reported:

Every loyal Old Town citizen points with pride to the Old Town Canoe Co., whose growth within a decade has been marvelous and the end is not yet, far from it. This year's product will amount to over 6,000 canoes, which cannot be handled economically with their present capability.

The people of Old Town awoke one morning and found half a dozen teams and about fifty men at work moving a long wooden building to Brunswick Street under the direction of Charles Roundy, and excavating for a brick building which is to be from 250–300 feet long and about 50 wide, so that when every building is built or relocated, they will have about double their present floor space. It is a busy place both inside and outside of this plant now and will be for some months. As soon as the wooden buildings are removed, more

9

A Growing Enterprise

definite plans as to the building will be made.

This concern is carrying the name of Old Town all over the world so that our city is getting free advertising that it could never afford to pay, from a dollar and cent standpoint. One advertising contract alone costs them about $15,000 per year giving us a little idea of their business. Their pay roll now is not excelled by any other plant in our city and is constantly increasing. They have recently purchased from the Union Land Co., their holding, between Middle St., and Stillwater Ave., consisting of a street and ten house lots where they are carting the dirt from the excavation for the new building. You need not be surprised to wake up some morning within the next year and find each and every lot covered with a modern residence.

The Old Town Canoe Co. is made up of Geo. A., Herbert, Samuel B. Gray, and George H. Richardson, a quartet of citizens who have had much to do with Old Town's present prosperity.

There used to be a slogan in this city some decades ago "watch Old Town grow," we would be more specific and say "watch the Old Town Canoe Co. grow." The next ten years will see a great transformation in the city of Old Town through their quiet, but effective efforts.

The most ambitious building, #4, was already taking shape. Three hundred feet long and five stories tall, it would epitomize the strength of the company. George again put his own crews to work. Like wheelbarrow-wielding ants, the men brought bricks up the ramp system for the next course. Two members of the crew were skilled Italian bricklayers, good friends who arrived together daily. But when the workday began, there was no time for friendship as each attempted to outdo the other, laying bricks as quickly as possible. With efforts like these, the building was completed in good time.

The new Building #4 was the jewel of the company and as the plant grew, a drawing of the cavernous buildings was included in the catalog, praising it as the largest canoe factory in the world.

The expansion provided the men with the space they needed to increase production and efficiency. Now construction became more specialized. Logs were brought to the sawmill by horse teams and cut to boards of the appropriate sizes. Clear, acceptable boards were selected and stacked to dry with spacers called stickers in between, which allowed the air to circulate. Wood that did not meet the quality standards for canoes was sold to the building trade or for other uses.

This artist's rendition in the 1915 catalog demonstrates the quick growth of the OLD TOWN CANOE COMPANY and clearly illustrates why OLD TOWN became known as the "Canoe City."
Gray Collections

In the yard, the lumber dried for a period of three months. Sitka spruce, which was purchased and shipped from the West, was so costly that it was brought immediately into the factory to lessen exposure to the elements. Because the yard was higher than the basement, the men used carts to move the wood over to an opening where they slid the lumber down into the basement, where all the woodworking machinery was located. Each operation had its own space, beginning in the basement. After the wood was planed or shaped, the workers moved the material to its appropriate department.

The offices were located on the first floor, right above the whirring noises of the saws and planers in the basement of Building #1.

Paddles and oars were made on the second floor. Rough-cut lumber was first cut to shape with a band-saw following a flat pattern; then, using the band saw again, the workers tapered each edge. The rough piece was then brought to the large 24-inch sanding drums, where the paddlemaker finished the paddle "by eye," sanding it to its elegant profile. The Penobscots were particularly adept at this art. They possessed the steady hands and light touch necessary for such an exacting job. After a few more turns on a lighter sanding belt, the paddle was sent to the paint shop on an upper floor for varnishing.

Also on the second floor, on the west side of the building, women were busy sewing. They made cushions out of kapok and covers of various kinds. There were no federal laws requiring personal flotation devices at this time, but OLD TOWN did offer the kapok cushion as a safety feature. Canoe lettering, fancy

painting, varnishing, and shellacking were completed on the third floor because it offered the best light.

Building #2 had no basement. The first floor was used for shipping. The canoes or boats were first wrapped in oiled paper. Burlap was laid on the floor and covered with hay to act as cushioning. The paper-wrapped craft were then placed on the hay and the burlap was drawn together and sewn shut. The wrapped canoes and boats were taken to the railroad yard by horse and wagon. The second floor of Building #2 housed all the accessories and repair parts that were needed or shipped to the customers. The third and fourth floors were used for storage.

Building #3 was used for very diverse purposes. The basement and first floors were used for storage, especially for skiffs, flat-bottomed boats, and later, lapstrake hulls. The basement later held an ash-

This Lancaster, Pennsylvania, merchant prominently displays his latest shipment of still-wrapped canoes.
Old Town Canoe Company

pounding machine. On April 25, 1925, the *Old Town Enterprise* was pleased to report on a new company product:

> The Old Town Canoe Company has a display of laundry baskets in the window of George Gray's Hardware Store. This is a new industry they have added to the canoe plant and we hope it will develop into a large and prosperous one.

This seemed like a natural application of the Penobscots' talents. Many of them were already caning seats and providing baskets for other markets. Now the canoe company would also offer their products. The ash-pounding machine expedited the basket-making process. A log was turned on a lathe while a flailing arm pummeled the log. The pounding softened the wood, weakening its fiber so that the workers could then pull away strips as they separated. Next, the

In the basement at OLD TOWN today, this now-antique equipment is still in use in the construction of wood-and-canvas canoes.
Susan Audette

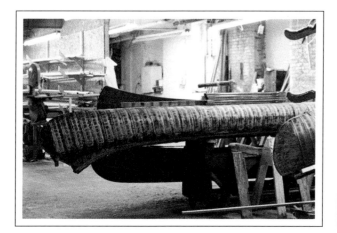

strips were cut to various widths, providing the splints needed for weaving. Many of the baskets were made at home and brought back to the factory. A wide selection was offered, made by skilled men and women, and this artistry and tradition is still carried on within the tribe today.

Painting and varnishing were restricted to the third and fourth floors of Building #3, away from the dust of the woodworking machines. Once the construction of Building #4 was complete—a building one story taller than #3—a ramp system allowed the men to exit Building #4 from its top floor and step directly onto the roof of Building #3. (The roof was railed to prevent anyone from falling off.) The canoes were often stacked five high, covering every foot of available roof space.

Machinery in Building #4 was again on the lowest level, and a blower/vacuum system had been installed to collect sawdust and scraps, which were used as additional fuel in the boiler.

The first floor was used for finishing canvas hulls

and in the late '20s for the construction of lapstrake hulls. The second floor was devoted to the first steps of construction—bending ribs on forms. This process was aided by the heating steam available throughout the building. The boiler pipes allowed the men to locate steamboxes conveniently for wood-bending operations. Six or seven pipes from the wood-fired boiler ran along the walls, providing heat for the cavernous plant. Although the heating system was state of the art for its time, the size and construction of the plant made it difficult for it to be truly effective. In the wintertime, if the men stood near the pipes, they were warm; out of the range of the radiant heat, they were freezing. (Of course, they dressed accordingly.) To increase overall efficiency, a ramp system was installed throughout the buildings so that as the men finished

Sam Gray was always looking for new markets, and the 1910 OLD TOWN CANOE catalog cover was designed to appeal to the independent woman.
Gray Collections

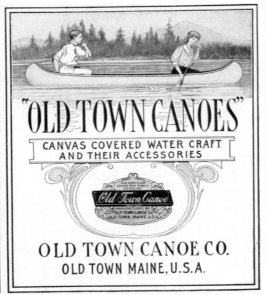

one operation, they could easily carry the canoe to an upper floor where its construction could continue.

The remaining floors of Building #4 were also specialized. The third floor was used for stretching and filling canvas on the hulls, the fourth floor for drying and storage, and the fifth floor for painting and varnishing. Every floor in each building had some space set aside for air drying ribs, which were stacked to the ceilings.

America's impending entry into World War I slowed things considerably at OLD TOWN CANOE. Sales for canoes —as for most types of sporting goods not associated with military life—took a sharp downturn during the war years. Not only were large numbers of men removed from civilian life, but for thousands of others the uncertainty of the situation made it seem unwise to make nonessential purchases. Sam continued to advertise with the same intensity as in previous years; even though customers might be scarce, it was not the time to stop looking for them.[1]

OLD TOWN CANOE's modern assembly line. Hoses attached to a blower/vacuum system removed shavings and sawdust from the canoes.
Gray Collections

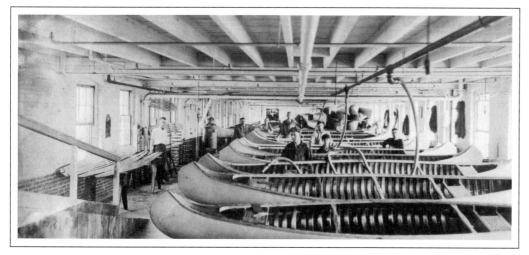

No history of Old Town would be complete without mentioning Linwood "Linny" Wickett, grand-nephew of Alfred E. Wickett. Linny started working for OLD TOWN CANOE in 1946. His title was "fireman," and he tended the fires and boilers at the factory. The fires were fed with scrap wood from the canoe factory and the Brewer Manufacturing Company, when that company was still in business. The boiler was used to provide heat for the factories and steam to bend wooden parts for the canoes.

Linny retired (sort of) in 1983 after thirty-seven years of service. He was given a 4-foot replica of an OLD TOWN canoe on a plaque as one of his parting gifts. He only "sort of retired" because Linny still returns to work two or three times a month to work on the boilers and burn the scrap wood. The managers say they still need him because he is their best boiler man. What is amazing about this is that Linny turned eighty-seven years young in August of 1998. He and his eighty-two-year-old wife, Alice, are still active farmers in West Old Town. They have sixteen head of polled Hereford cattle, tend a small garden, and make maple syrup in the spring. Linny still cuts, rakes and bales most of his own hay, and Alice drives the truck for the neighborhood crew to load the hay to take it to the barn.

Linny has a long list of funny stories to tell, but the following story is about him. Linny's chair still stands in the OLD TOWN boiler room today. That chair is almost as old as his tenure at OLD TOWN, and Linny has accumulated several years' time in that old chair, always keeping a watchful eye open on the fires that keep the factory running. In years past, he spent many nights on the night shift, and surely he stayed alert, except one time.

Linny had apparently spent a long night at home the night before work, perhaps helping a cow give birth or chasing "Wifey" around the kitchen table. Linny dozed off in his comfy chair, and several workers gathered to view the sleeper. One of his co-workers thought he'd be funny and started spraying a hose around the chair, seeing how close he could come before Linny woke up. A little water sprayed on him, and Linny woke up with hands flailing and feet flying. His boss, Jimmy Cunningham, caught the action and said, "Get him, Linny!" So Linny chased the prankster out the fireroom door, giving him a wet tail end!

—Linwood B. Wickett, Jr., Stratham, NH

On the contrary, he had to search all the harder. The advantages of canoeing could still be pointed out to those at home who had to keep fit in order to meet the nation's jobs. Sam also recognized that more women were working than ever before and made much better wages. These women were a new potential market for OLD TOWN canoes, and he used advertising designed to appeal to them.

Because of the fluctuating market, the 1917 catalog was a bit late, and its prices reflected increases in the prices of raw materials also used in the war effort. An overall price increase of $3 per canoe or boat was noted, brass bangplates for the full length of the keel rose 50¢, and copper air tanks were $5. The supply of raw materials was not the only problem. Congestion on the eastern railroads made delivery to dealers difficult. Like so many other cities, Old Town was suffering from a scarcity of incoming fuel, affecting OLD TOWN's ability to heat their buildings. But Sam Gray would not let the adversity get him down. He continued to keep the OLD TOWN name in front of the public, reassuring his dealers while he vigorously pursued new markets.

At home the OLD TOWN CANOE COMPANY supported the war effort by making paddles and oars that were used on assault boats, lifeboats, and rafts. The company, along with the Gray family, gave generously during the period to the Red Cross and to bond campaigns. Even the eldest of Sam Gray's children, son Braley, remembers knitting socks to help the troops when he was just a boy. Everyone did their part.

Fully entrenched in the war effort, the company reflected the patriotism of the country and used it to its best advantage. The 1919 catalog depicts a soldier with rifle and bayonet drawn. "Outdoor Life Did It," the cover read, crediting his mettle to his outdoor spirit and an OLD TOWN canoe. A stamp attached to the cover alerted customers to an advance in price, effective April 25, 1919, of 15 percent. Inside the cover was a patriotic piece and three new paint options were available, all camouflage style, presented in color for the first time: a maritime pattern, greenbank, and tiger patterns. The catalog also showed specifications in chart form for the first time.

OLD TOWN had already made a strong entrance into the sportfishing market in 1917 with the introduction of a square-sterned model when "detachable motors" began to play a larger role in recreational boating. As always, the company was quick to respond to new trends, with power tenders to replace the dinghies, and racing boats for those looking for a challenge or thrill. Powerboating and yachting magazines now carried OLD TOWN ads. By 1923 OLD TOWN CANOE took on the distributorship of the complete line of popular Johnson outboard motors, becoming Johnson's first distributor.

The earlier acquisition of the Carleton line had added a new dimension to OLD TOWN. With Carleton's capabilities, OLD TOWN could vary its products without having to alter its own operation. It could also sell to rival retailers by offering the two different lines. In New York City two large department stores offered canoes: Macy's bought OLD TOWN and Gimbel's, the Carleton line. As the years progressed, the operations of the two canoe companies overlapped. Merchandisers may have realized that they were buying the

The OLD TOWN CANOE *catalog cover of 1919 showed the tenor of the times.*
Gray Collections

Far right: The introduction to the 1919 catalog credits the stamina of our men in battle to the outdoor life, implying that canoeing may have played a part.
Gray Collections

Old Town Canoes

"Outdoor Life Did It!"

These patterns were anything but camouflage but were reflective of the times. Gray Collections

Far right: Introduced in the 1917 catalog, the Square Stern Canoe was developed to meet "the widespread use of portable engines.... To it can be clamped any of the standard makes of Rowboat Motors." The Square Stern was equipped with cross seats, but these could be omitted and two folding cane chairs substituted. A further option was the installation of sponsons. This Square Stern was photographed in the 1950s. Gray Collections

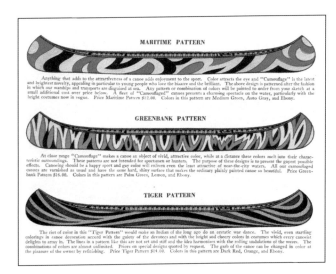

MARITIME PATTERN

Anything that adds to the attractiveness of a canoe adds enjoyment to the sport. Color attracts the eye and "Camouflage" is the latest and brightest novelty, appealing in particular to young people who love the bizarre and the brilliant. The above design is patterned after the fashion in which our warships and transports are disguised at sea. Any pattern or combination of colors will be painted to order from your sketch at a small additional cost over price below. A fleet of "Camouflaged" canoes presents a charming spectacle on the water, particularly with the bright costumes now in vogue. Price Maritime Pattern $12.00. Colors in this pattern are Medium Green, Auto Gray, and Ebony.

GREENBANK PATTERN

At close range "Camouflage" makes a canoe an object of vivid, attractive color, while at a distance these colors melt into their characteristic surroundings. These patterns are not intended for sportsmen or hunters. The purpose of these designs is to present the gayest possible effects. Canoeing should be a happy sport and gay color will enliven even the least attractive of near-the-city waters. All our camouflaged canoes are varnished as usual and have the same hard, shiny surface that makes the ordinary plainly painted canoe so beautiful. Price Greenbank Pattern $16.00. Colors in this pattern are Palm Green, Lemon, and Ebony.

TIGER PATTERN

The riot of color in this "Tiger Pattern" would make an Indian of the long ago do an ecstatic war dance. The vivid, even startling colorings in canoe decoration accord with the gaiety of the devotees and with the bright and cheery colors in costumes which every canoeist delights to array in. The lines in a pattern like this are not set and stiff and the idea harmonizes with the rolling undulations of the waves. The combinations of colors are almost unlimited. Prices on special designs quoted by request. The garb of the canoe can be changed in color at the pleasure of the owner by refinishing. Price Tiger Pattern $14.00. Colors in this pattern are Dark Red, Orange, and Ebony.

After enjoying a very pleasant day on the lake, my husband was backing the boat trailer down the boat-launch ramp with our pickup truck to retrieve our OLD TOWN Square Stern model. While I was holding the boat dockside, we attracted the attention of a six-year-old boy and his younger brother, who were standing near the breakwater opposite me. The six year old decided that it was his duty to narrate the proceedings and began loudly, so everyone around could hear, including my husband.

First, it was, "Look, Hermie, he's gonna go get that boat. Watch him. Ya see that, Hermie?"

Then, as my husband began to back into the water, we heard, "Look at that, Hermie. It's getting really wet!"

The tone of his voice was becoming more and more animated as he kept repeating, "See, it's goin' in deeper, Hermie, it's getting deeper!"

He heard the burble from the exhaust pipe in the lake and saw that the trailer was almost completely underwater.. Apparently thinking my husband was going to back the truck all the way into the lake, he just couldn't stand it any longer and issued the full alarm. "Mister, Mister," he hollered, scaring the daylights out of my husband, who slammed on the brakes, afraid something was seriously wrong.

Even louder yet, we heard, "Mister, Mister, roll up your windows!"

My husband got laughing so hard he just shut the truck off until he could compose himself.

—M. J. Baker, Oaksville, NY

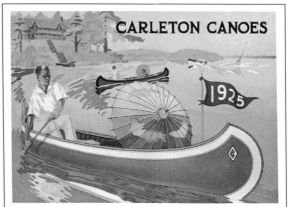

Carleton Canoe catalogs from the 1920s claimed: "The Carleton Model Canvas Covered Canoe is a reproduction from a birch bark canoe belonging to the Penobscot or Tarratine tribe of Indians. It has been on the market for more than thirty years, being the pioneer in the modern canvas canoe. During all this period the continuing commendation of canoe users in general, including professional guides and Indians themselves, has kept us firm in the belief that our model in every detail is perfection and it has remained unchanged."

Courtesy of Jim Cunningham

Specification charts and diagrams were helpful to prospective buyers. This arrangement appeared in 1920 and subsequent catalogs.
Gray Collections

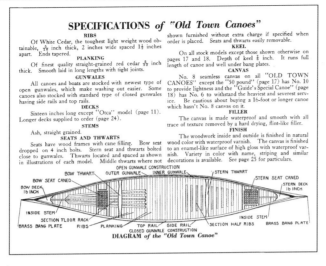

SPECIFICATIONS of "Old Town Canoes"

RIBS

Of White Cedar, the toughest light weight wood obtainable, ⅜ inch thick, 2 inches wide spaced 1½ inches apart. Ends tapered.

PLANKING

Of finest quality straight-grained red cedar ³/₁₆ inch thick. Smooth laid in long lengths with tight joints.

GUNWALES

All canoes and boats are stocked with newest type of open gunwales, which make washing out easier. Some canoes also stocked with standard type of closed gunwales having side rails and top rails.

DECKS

Sixteen inches long except "Otca" model (page 11). Longer decks supplied to order (page 24).

STEMS

Ash, straight grained.

SEATS AND THWARTS

Seats have wood frames with cane filling. Bow seat dropped on 4 inch bolts. Stern seat and thwarts bolted close to gunwales. Thwarts located and spaced as shown in illustrations of each model. Middle thwarts where not

shown furnished without extra charge if specified when order is placed. Seats and thwarts easily removable.

KEEL

On all stock models except those shown otherwise on pages 17 and 18. Depth of keel ⅞ inch. It runs full length of canoe and well under bang plates.

CANVAS

No. 8 seamless canvas on all "OLD TOWN CANOES" except the "50 pound" (page 17) has No. 10 to provide lightness and the "Guide's Special Canoe" (page 18) has No. 6 to withstand the heaviest and severest service. Be cautious about buying a 16-foot or longer canoe which hasn't No. 8 canvas on it.

FILLER

The canvas is made waterproof and smooth with all trace of texture removed by a hard drying, flint-like filler.

FINISH

The woodwork inside and outside is finished in natural wood color with waterproof varnish. The canvas is finished to an enamel-like surface of high gloss with waterproof varnish. Variety in color with name, striping and similar decorations is available. See page 25 for particulars.

DIAGRAM of the "Old Town Canoe"

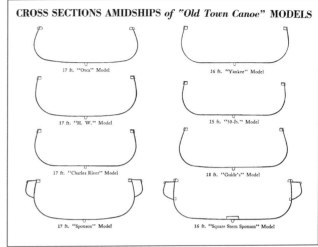

CROSS SECTIONS AMIDSHIPS of "Old Town Canoe" MODELS

17 ft. "Otca" Model

16 ft. "Yankee" Model

17 ft. "H. W." Model

15 ft. "50-lb." Model

17 ft. "Charles River" Model

18 ft. "Guide's" Model

17 ft. "Sponson" Model

16 ft. "Square Stern Sponson" Model

"A thrill for speed exhilarates the wistful observer of this boat. Designed by the engineers of Johnson Motor Co., it embodies lines that make water racing seem as fast as flying. Rated at 16 miles per hour with Johnson Big Six Motor—and some owners have reported even better speeds." From 1927 catalog copy for the 16-foot OLD TOWN "Baby Buzz."
Gray Collections

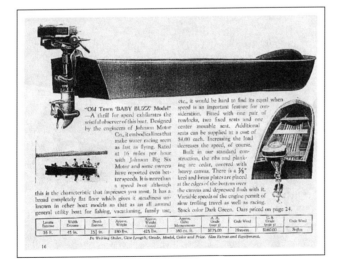

same product with different labels, but no one in the general public seemed to mind. Retailing greats like Abercrombie and Fitch of New York City and Iver Johnson of Boston ordered OLD TOWN canoes by the railroad carload.

When the war ended, there was a new surge in recreation and Sam Gray continued to use his marketing savvy to enthusiastically welcome the troops home. Production at OLD TOWN was at a frenzy to meet new orders, and attempts had to be made to increase production.

Ralph Brown, a college graduate, had been hired to improve efficiency and came in to assist after Wickett's departure. Brown was responsible for many innovations at the factory, including the blower-vacuum system in the basement and ventilators and exhaust fans in the color rooms. The locals had been amused when a college graduate was hired to implement changes that they felt anyone could see were necessary. There was also some resentment when this "newcomer" attempted to tell some of the old-timers how to do their jobs. Brown even recommended changes in the Guide model, a long-standing design that had a great following among those who made their living in the woods. He changed the straight bow and stern into "impossible curves" that brought howls of protest from the guides themselves. The wind would catch in those curved ends, and they were generally thought to be a nuisance. The company lost no time in getting the straight, utilitarian ends back on the Guide.

Brown also attempted to speed up production. His target was the filling and drying process. After a canoe's canvas was coated with lead-based filler, it needed to dry slowly for four or five days before the next coat could be applied. Then another three to five days drying time was needed for the second coat, followed by weeks of curing. Brown recommended drying bins much like those used in producing ceramics. Bins, much like kilns, were built and fitted with steam pipes to supply heat. With this new system, it took only four days to dry a canoe. Sam Gray was so pleased with Brown's invention that there was talk of Brown being appointed general manager, in addition to receiving a few blocks of shares. His innovation came just in time to complete an order of 500 canoes for the Macy Company of New York City. The workers broke all records in getting the canoes out—and Brown's bins made it possible.

The success was short lived. About two months after the last canoe was shipped, the canoes began to come back. Many were warped and had cracked ribs

and splintered gunwales. Some of the canoes had places "where big gobs of paint, a foot across had dropped off."[2] Of the 500 canoes ordered, 275 of them came back to be repaired or replaced. Needless to say, Brown quickly found he was no longer on OLD TOWN's payroll.[3]

Sales soared in the years following the war, and advertising continued to beckon new buyers. In 1927 $25,000 was budgeted for advertising, almost doubling the amount spent just ten years before. The increase paid off as sales passed the half-million dollar mark.[4]

On April 24, 1928, George Gray, Sam's father and mentor, died. His death left not only a family in mourning, but a city as well. The day before George Gray's funeral, a proclamation was issued by John Hickey, mayor of Old Town, stating:

> Inasmuch as the late George Gray, a citizen of Old Town for over 82 years, by his tenacity of purpose, honorable dealings, and keen judgment, built up businesses which have aided and are aiding this city toward that measure of prosperity which reflects itself in the well-being of its citizens, and inasmuch as he was a generous and kindly man who quietly performed many acts of help toward others...that in respect to his memory all places of business herein be closed...Saturday morning...at the time of his funeral.

It was an unprecedented gesture. Old Town was quiet on the morning of Saturday, April 28, 1928, as everyone thought of George Gray. A founding father was gone but certainly not forgotten.

The Salesman's Samples

We all like to get a little extra with our purchases, and Sam Gray was quick to recognize this. He rewarded companies who bought large quantities of canoes with a model canoe, a "salesman's sample" in today's vernacular—a beautiful miniature canoe that could be put on display. The hand-crafted, wood-and-canvas canoe was the exact replica of the larger models. Several of OLD TOWN's workers built them at home in their spare time and were paid extra for their efforts.

Buyers received a 4-foot model if they ordered enough canoes to fill one railroad car; if they filled two, they had their choice of an 8-foot model or two of the 4-footers. These precious miniatures are rare today and demand a high price. In auctions over the past few years, the canoes have brought from $3,000 to $12,000, a high price for a piece of the past.

Recognizing the value of these old salesman's samples, OLD TOWN CANOE recently restored one of the originals that hangs in its office area. In addition, OLD TOWN commissioned Jerry Stelmok, a premier canoe builder and restorer in Atkinson, Maine, to make several of the small models to offer in the OLD TOWN catalog; they are priced at $2,500 for a 4-foot model, and $2,900 for an 8-footer.

"Old Town 'CHARLES RIVER MODEL' Canoe"

"Old Town 'YANKEE MODEL' Canoe"

Color Design No. 2, $10.00

Color Design No. 4, $18.00

Color Design No. 7, $16.50

Color Design No. 20, $3.50

COLOR DESIGNS

In these end sections of "OLD TOWN CANOES" are shown a wide range of designs for the whole length of the canoe. In ordering please specify the design number and price as indicated. These designs are susceptible of various color schemes, and can be executed in any combination of colors you may submit. Price of design No. 23 includes mahogany rub rails ($5.00) which separates the colors. Designs on the catalog cover are priced on page 25. This full size canoe shown below is Color Design No. 29, $16.50.

Color Design No. 21, $10.00

14

Color Design No. 22, $16.50

Color Design No. 23, $14.00

Color Design No. 25, $5.00

Color Design No. 26, $10.00

15

Color Design No. 27, $11.00

When I was in sixth grade (1964), my father's friend offered him a free canoe, which turned out to be a 16-foot 1913 OLD TOWN Charles River model (serial #22091). It was in such bad shape that we brought it home in the dark so as not to offend the neighbors.

—R. Rybinski, Manlius, NY

In 1958, my wife, our three kids, and I were in the Boundary Waters above Ely, Minnesota. My wife and I had an old handmade canoe and carried our daughter as a passenger. Our two boys, ages ten and twelve, were using the 1936 OLD TOWN wood-canvas "Yankee" canoe. About four days into the bush, we were making a portage, carrying our stuff across, when along came the two boys but no canoe! They had had an argument and were in a big pout.

I asked, "Where is the canoe?"

They said, "We threw the damn thing away!"

I said, "How will you get home?"

They said, "We don't care." We went back down the trail and found the canoe lying in the bushes. From that point on, it was my wife and one boy and me with the other. Our daughter was, of course, a good passenger.

—Carl L. Hoth, Sun City, AZ

My wife and I had rented a cottage only about 50 feet from the shoreline of Blue Mountain Lake in the Adirondacks. At first light one morning, we were sitting on the porch, sipping coffee, waiting out a light rain shower on the lake. We wanted to take a quiet paddle in our 1916 OLD TOWN Charles River canoe before breakfast. It was almost ethereal; everything was hushed and shrouded by the rain and mist. Our canoe waited patiently, having been turned upside down on the beach between the lake and us. So as not to break the mood, we only occasionally whispered to each other.

Before long, two kids—a boy about nine years old and his younger sister of about seven—came along. They were dressed in rain slickers and each was carrying a fishing pole. The boy had a small tackle box. They, too, were caught up in the mood and talked very quietly between themselves but just loud enough for us to hear them. Not detecting us at all, they walked out to the small dock adjacent to our beach and began fishing. It soon became obvious that the boy was giving his sister her first fishing lesson, and also that the dock wasn't large enough for both of them to occupy it at the same time. We were delighted by their demeanor and innocent caring for each other. Of course, the boy had to try to fling his plastic, chartreuse with red stingers, "catch-em-all" into the deep part of the lake, but he took the time to instruct his sister how to do it, too. She would stand back and fish the shallows while he gave it a try, and then he would invite her up for a turn, changing places with her. We were held spellbound by their behavior. They never raised their voices, but they did talk continuously. I guess first fishing lessons probably demand that. Finally, after this went on for close to half an hour, the inevitable happened and the "catch-em-all" got hooked on the bottom. Uh oh! They pulled, tugged, and each offered every suggestion for how to get it loose, without raising their voices, but nothing worked, and eventually it was lost. The boy was obviously upset, but his sister smoothed everything over by saying, "It's still pretty early. I guess the fish aren't up yet." He agreed, packed up their gear, and they went back to their cottage, quietly. It had just stopped drizzling, so we went for a paddle, quietly.

—Dave Baker, Fly Creek, NY

1   "Old Town Canoes Advertised as Usual, Despite Smaller Sales," *Printers Ink* (1918), p. 111.

2   Robert Grady, " The Life of Henry Mitchell," *Living Lore in New England* (WRPA Project: Oral History 1938–1939), p. 7.

3   *Ibid.*, p. 7–8.

4   Gray, "How We Built New Markets," *System*, p. 68.

Sam was deeply saddened by his father's death. As Sam's mentor and friend, George had been responsible for the man Sam had grown to be. Sam was grateful and expressed this in the company's log. Through the difficult times ahead, Sam began to rely on one man who also had often been at his side— Pearl Cunningham.

Cunningham was originally from Patten, Maine, and came to work at the hardware store as early as 1900. He doubled up his sales responsibility in the early years by going on the road for Gray's Hardware and for the Bickmore Company. His sales trips brought him and his team of horses as far west as Iowa. Cunningham, like the Grays, valued a great horse. Few people had the experience with horses and, more important, with managing people, that Pearl possessed. It was natural for George Gray to ask Pearl to extend his duties to include sales for the canoe company.

As other superintendents came and left, Pearl Cunningham moved closer to the top, first becoming the sales manager and finally by 1911, the general superintendent of the OLD TOWN CANOE COMPANY. His move to the number two position may have been one of the factors that forced Wickett's decision to move on.

Cunningham had proven himself a dependable individual to both George and Sam Gray. Small and energetic, Pearl possessed an uncanny attention to detail that extended beyond sales orders into every corner of the factory. His notes included records on production, right down to the amount of wood needed to operate the boiler. For every note that was written on paper,

10

Relying on Pearl Cunningham

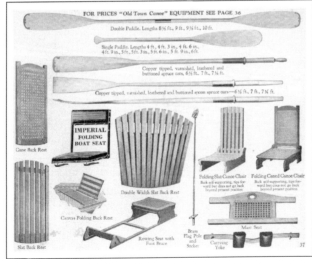

there were volumes that were not. Perhaps Pearl knew that this was one way to make himself irreplaceable, and he kept much of the operation in his head. The combination of Sam Gray's astute business sense and Cunningham's reliability and knowledge of all aspects of the company was ideal, and they developed a relationship that would carry on for several decades. To Sam, Pearl was irreplaceable. In the factory and in the woods, the men considered Pearl tough but fair. He was respected by the workforce and took time to know the men personally. It was not unusual for Pearl to bail someone out of a jam—even jail—or to advance a week's pay to a man in need.

Despite all the demands of running the plant, Cunningham continued his sales trips west through the '20s, leaving in September and returning with orders in hand just before the Christmas holidays. It was a routine that his family knew well. His December 27, 1924, sales log reported: "Results of fall selling trip include: Old Towns 556, Carletons 256, Paddles 3,417, Double-Bladed Paddles 161, #2 (blem-paddles) 2,170."

During the '30s sales were erratic. In a cost-saving measure Carleton, which had been maintained as a separate company, was consolidated under the OLD TOWN name on December 18, 1934. However, the company continued to issue a separate Carleton catalog. The Depression hit OLD TOWN hard, with business dropping off to a fraction of what it once was, and the company struggled to stay alive. Sam's theory was, "This damned Depression can't last forever." Together with Cunningham, he tried to keep the men working at anything they could, repairing or adding onto the factory, or cutting wood for the furnaces.

Sam was right. The Depression did not last forever, and the company saw a resurgence of interest. With an improving economy, orders were again picking up. On his many trips, Pearl not only took orders, but he also purchased the horses needed for the logging operations and for work around the yards at the canoe factory.

The logging operations varied over the years. OLD TOWN CANOE had begun importing Western red cedar for planking by the 1920s because it was difficult to find sufficient clear cedar stock in Maine. The trees coming from the West Coast were often so large that they had to be quartered before they could even be put on a bandsaw. Some years OLD TOWN CANOE could rely on other lumbermen to supply its needs, but during the Depression and war years, it resumed its own operations more aggressively. In the early '40s the woods operations that supplied much of the necessary wood for the canoes and boats and boiler operations at the factory increased to five camps under Pearl's watchful eye, with camps as far away as Dickey, near the St. John River.

The pressure of World War II affected everyone in town. Many of the Penobscots who had previously worked in the canoe factory now found more lucrative opportunities elsewhere. The able men of Indian Island enlisted, achieving 100 percent participation.[1] Some left the canoe factory for higher paying defense jobs.

Because of enlistments, there were fewer men working in lumbering; raw materials would only be available if OLD TOWN sent its own men to the woods.

With OLD TOWN workers in short supply, employees often did double duty, some working in the woods as well as in the canoe factory. When work was slow, Sam Gray even put them to work on repairing the houses he owned or his summer camp.

Again, OLD TOWN CANOE was busy supplying ash oars and paddles for the new war effort and by doing so, was able to continue building canoes and boats, but in fewer numbers. The government provided fuel

Lumbering activities changed over the years. Here, river drivers use an OLD TOWN strip-planked boat (left) to move the pulpwood.
Old Town Canoe Company

allotments so OLD TOWN's operations could continue, and a sawmill was set up in Building #6 to handle the wood from the camps. Sam Gray was approached by the army to build an assault boat, complete with plywood pontoons. The army supplied a series of blueprints with the intricate detail they required. After Cunningham reviewed the plans with his son Jim, a mechanical engineer working in New Jersey, Pearl and Sam agreed that the design was too cumbersome for their facility. Instead they concentrated on supplying paddles, oars, lifejackets, and box cushions to the military, even though materials were difficult to obtain.

The war years took their toll on the canoe business. Brass and other precious metals were reserved for military contracts only. In their place, the men used steel nails and galvanized steel bang strips. Kennebec, a canoe company in Waterville, Maine, wrote OLD TOWN wanting to buy a keg of nails. OLD TOWN replied that it "had less than ten days' supply on hand" and was not able to get any more from its supplier.[2] OLD TOWN would, however, ship some nails after it replenished its own stock.

Canvas was also in meager supply. Sam Gray and Jim Cunningham had petitioned the government to lift restrictions so they could continue manufacturing canvas-covered canoes and boats. At one point they were contacted by a man in Portland who had a supply of canvas that would build six canoes. Sam was happy to put some of his men to work, and when the canoes were ready, the buyer sent a truck to pick them up. The canoes were wrapped in the standard fashion, first in oiled paper, then in burlap cushioned with hay.

Using crosspieces, the workers stacked the canoes on the back of the flatbed. Not long after leaving the factory, heading down the road outside Bangor, the driver lit his cigarette and mindlessly threw the match out the window. Within a few minutes his cargo was ablaze. By the time he realized what had happened, the flatbed was a traveling pyre, and it was too late to save the precious cargo. As he looked at the charred skeletal remains of the six canoes, he wished that he were overseas with the others who were serving.[3]

By the mid-1940s, OLD TOWN had five logging camps in full swing, all located within the Penobscot River basin. Camp 1, located in Township 39 about 25 miles to the east, had its own mill and specialized in cutting just cedar. Camp 2 was the Katahdin Iron-works, 40 miles north. The Ebeme Lake and Knight's Landing camps were in the same general area. The final camp, near Smyrna Mills, was run by Guy Friel, a local logger, and his sons. They had their own mill and supplied several crews. When necessary, timberland was leased. It was Pearl's responsibility to cruise the forest, looking for just the right mix of pine, ash, oak, spruce, and cedar. Once a desirable area was found, the land was leased. Owners were paid stumpage for the standing timber that was cut.

Under Pearl Cunningham's direction, the woods operations began in the fall with each person's role clearly defined. Roads were laid and prepared as well as they could be, and the camp would be centered in a lumber stand. All camps had a crew of about fifty men. There were teamsters to work the horses, cutters for sawlogs and pulp, a cook, and cook's helper. The cook was often under-appreciated, and there seemed to be an ongoing war between him and the crew, likely due to his waking them each morning by banging on a frying pan and yelling, "Roll-out!"[4] A clerk did the paperwork, the payroll, and often attended to the men's medical needs. The boss did the hiring and firing, laid out the work, and stayed at camp to oversee the operation. Because this job was so important, camp bosses were selected by previous work records or recommendation. They usually had their own preferences for clerks and cooks—and, sometimes, even some cutters.

Bosses often arrived with their own experienced men who erected all the necessary buildings. Once an appropriate site was chosen, the men built bunkhouses and outhouses, an office building, hovels for the horses, and a nearby yard or wood storage area. A blacksmith was needed to tend the horses and repair axes, canthooks, pick poles, Peaveys, and other various pieces of equipment. The blacksmith would also forge shoes for the log-hauling sleds, which were crucial to the operation.

Once the weather turned cold and the ground froze, the men started cutting. Usually, two cutters worked with a teamster, whose horses were key to their success. Not everyone could handle the horses properly, and it was no secret that an OLD TOWN teamster would be fired in an instant for any mistreatment of the animals. The horses would skid the logs from the woods to the landing, where sleds would be loaded with as many as fifteen logs stacked butt end first on a sled, then hauled to more centrally located yards. Some of the collection yards were large, often with up to 100,000 board feet of logs. The logs were sorted, rolled onto cribwork that had been constructed at the height of a truck bed, loaded onto a truck, and taken to the sawmill.

Occasionally the horses dragged the logs out to the side of a main road to be trucked later to the sawmill, usually near the main camp. Because trees were cut by selection and not clear-cut, horses were the most effective means of moving the logs in the woods. Tractors were not used. As the years progressed, tractors were used for breaking roads and hauling big skids of logs, but horses were employed in some aspects of OLD TOWN's operations right through the 1950s.

Because they were paid for what they cut, woodsmen learned to use their tools efficiently and looked for ways to improve them. As the logs were moved out of the woods, the loggers used canthooks to make rolling them easier. A canthook is a stout wooden lever with a "dog," a removable metal arm with a sharp hook at the end. Joseph Peavey, an ingenious local blacksmith from nearby Stillwater, replaced the staple-fastened dog of the early canthooks with a collar-clasped dog and added a spike at the end of the pole, greatly improving its efficiency. The tool has everafter been known as a "Peavey," or "cant dog."

Logging was dangerous work. Safety was paramount, so operational decisions were made prudently. A miscall could jeopardize the lives of men and horses. Work often continued into late April until thawing, sloppy conditions made it too difficult or risky to continue. Occasional accidents occurred as teamsters pressed their luck, but only once did OLD TOWN lose a horse. From Pearl's log: "Saturday, Feb.

21, 1925. Rained tonight. Sleds broke through the ice today. Had horses hooked to end of pole so unhooked them off. 2 hooks of ours and one of Hayden's went thru.... Get off the river now and stay off!"

In the spring when the camps closed, logs were trucked back to the factory and stored in a yard between Building #6 and the Murphy barn where the horses were housed. There the logs were stored until they could be sawn. After sawing, the lumber was graded for quality and taken to yards around the factory to dry. Wood intended for factory use stayed in the company yard. Other wood not suitable for factory use was dried—sometimes in Cunningham's backyard a few blocks from the plant—and later sold. The wanigan, a supply house on wheels with a collection of camp gear that could be reused yearly, was stored at the company. The horses were sold and repurchased annually to cut down on the costs of feeding and housing them through the summer.

As if overseeing the logging operations was not enough, Pearl Garfield Cunningham was also a key part of the "research and development" team at OLD TOWN, before the phrase was ever coined. He tested new products and made notes about any change in performance prior to that change being adopted. From the woods operations and the mills to the canoe manufacturing and shipping, Pearl Cunningham knew it and did it well.

Christmas was a time that signaled joy in the Cunningham family—it was the time when Pearl assured them he'd be home. He always made it back for that important day with his family. He was home that December in 1945, but he never got up Christmas morning. Without any forewarning, he had died in his sleep.

1 Ted Mitchell Interview, University of Maine, August 12, 1992.
2 Kennebec Canoe Collection, Maine State Museum, Box 3, 90.42.40.
3 Deane Gray Interview.
4 "Lumber Camps of Many Years Ago," The *Penobscot Times*, March 27, 1941, np.

In 1930, college student Bill Workman and three other waterfront staffers at a Boy Scout camp near Chattanooga got the idea to mount a wilderness expedition to paddle two of the camp's OLD TOWN canoes through the Hiwassee River gorge, from Murphy, North Carolina, all the way back out to Chattanooga. After the camp was over that summer, the four young men built wooden crates for the two canoes and hauled them down to the L & N Railwood Depot in Chattanooga, shipping them by rail to Murphy, North Carolina.

Their river route covered 133 miles, 98 miles down the Hiwassee until it emptied into the Tennessee River, and then 35 miles down the Tennessee to the Walnut Street Bridge in downtown Chattanooga. This was then free-flowing water all the way, six to ten years before the Hiwassee, Appalachian, and Chicamauga dams were started by TVA. Today, only about ten miles of this route is free-flowing gorge whitewater, and then only when the TVA powerhouse generators are discharging. The rest is dry gorge, slow-moving flat river, or calm lake. The difference is called progress, or electricity.

The total drop in elevation is 866 feet, from the L & N Depot in Murphy, to downtown Chattanooga. Most of that elevation drop occurs in the first fifty miles on the upper half of the Hiwassee. Today, one 12-mile section is normally dry river bottom gorge where water is piped underground to its powerhouse. After high rainfall, on rare occasions, TVA releases water down the gorge, unannounced ahead of time. Only a few people have caught a release and paddled this stretch since the dams were built.

This 12-mile section was the most challenging section to our 1930 adventurers. The river averages a drop of 28+ feet per mile! In one of those miles (river mile 58 to 57), the river drops approximately a hundred feet, beginning with the highest falls, now named "Workman Rapid" in honor of Bill Workman, James Irvine, Luke Lee, George Stewart, and the two loaded wooden OLD TOWN canoes that ran it in 1930.

The four adventurers loaded the two OLD TOWNs with pup tents and provisions for about two weeks. They camped beside the river each night. Each bowman had a 10-foot-long pole to fend the canoe off rocks in the river, and some of the photographs from the trip show a canoe upside-down at a campsite, titled "Canoe repair." Other shots show canoes being lined down rapids and other riverside campsites. Missing in all the photos are life-jackets!

As far as we know, this was the first recorded descent of the Hiwassee River, at least by white men, and also the last descent before the main rivers of the Tennessee River Valley were dammed for electricity during the '30s and '40s. We don't know what eventually happened to those two OLD TOWN canoes. For many years, they were stored in Captain Gladish's boathouse on the Tennessee River, across from what is now the Tennessee Aquarium. They were used for recreation by Bill Workman and other Scouters, and his wife, Betty Peruchi Workman, sewed sails for them.

In 1956 Bill proudly bought a brand-new dark green 18-foot OLD TOWN Guide, virtually identical to the ones he paddled in 1930, and taught his two teenage sons and me, his nephew, the lure and lore of flatwater canoeing. Bill passed away in 1989, and the restored 1956 Guide is now my proudest possession.

—Jack Wright, Chattanooga, TN

11

A New Generation

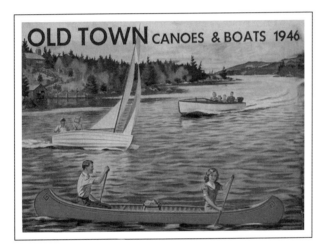

P earl's death left the workers at OLD TOWN stunned and the town was equally dazed. The Grays mourned their friend but they had to deal with the reality of continuing with the day-to-day business of running a canoe factory. Pearl was virtually irreplaceable, and although there might have been local skill to reduce the gap, Sam Gray was taking no chances on totally unknown talent. Over the years Braley, Sam's oldest son, had accompanied Pearl to boat shows throughout the country and since 1936 had worked at the plant, assuming many of Pearl's responsibilities in an effort to share the load. It wasn't difficult for him to step in and continue most of the sales work. His affable personality made him a natural for the duties and the position. However, there was much more to Pearl's job than sales.

Sam's youngest son Deane had recently completed his tour of duty with the Army Air Corps. Realizing his father needed help, he put other interests aside and came back to the factory to pitch in. Because much of what Pearl Cunningham had done had been recorded only in his head, the company needed new accounting methods and a complete overhaul of an antiquated inventory system. Deane took this on as his personal task. The only areas of the factory affairs left unsupervised were the logging operations, the firerooms, and the machinery—not easy areas to oversee.

The day after Cunningham's funeral, Sam Gray asked Jim, Pearl's oldest son, to visit him at the factory. Jim was no stranger to the plant. During the summers he had worked in the paint shop, earning thirty cents an hour for varnishing and shellacking paddles. Later he worked nights making flat-bottomed boats—good employment for a young man, close to home, yet under his father's watchful care. His fondest memories of the company were as a child, when the factory was his playhouse and the workers encouraged his visits. His favorite place was the basement, where the Indians would often treat him to a small bow and arrow made from floor scraps. With his bow and arrow, the young boy would protect his empire, ready to take on any offender. Reality returned abruptly one day when an errant arrow pierced an evil-doer, which turned out to be a newly canvassed canoe. Without hesitation Pearl knew whom to blame, the remorseful Jim was sent home, and the arrow became kindling.

Following the funeral services for his father, Jim had intended to return to his mechanical engineering job at Wright Aeronautical Company in Paterson, New Jersey, but Sam Gray offered Jim a salary and a steady job carrying on Pearl's work at whatever was needed. With the uncertainty of employment at Wright's, Jim decided to accept the offer and started work on January 15, 1946. The winter season dictated that he devote his attention to overseeing the logging operations, and he had much to learn and understand. Like his father had done before him, Jim traveled about 200 miles per day to supply the camps' needs: men, food, medicine, blankets, fuel, tools, hardware, horses, hay, and grain—all part of the supply line to keep the camps up and running. He kept daily records of cuttings and got to know the men involved in the camps, as well as the people and companies that supplied their needs.

After mastering the logging operations, Jim's skill

was applied to other areas. Before long he was designing machines to increase production. He first completed a machine designed to automatically put the taper on ribs that were used in the canoes. This operation added more uniformity to the finished product. In addition he designed a special clamp for use in the construction of strip-planked hulls. Over the years he rewired the entire factory, redesigned a complete new heating system, and installed a new oil-burning furnace. The aging building provided many challenges, but Cunningham had the engineering background to bring new ideas to reality.

The government's requirements during World War II had created more than just shortages. New regula-

Braley Gray learns another aspect of the business as he works on completing the planking on a wood hull.
Courtesy of Braley Gray

tions were placed on labor practices. OLD TOWN CANOE COMPANY, for the most part, paid its employees for piecework, but now were forced to offer an hourly wage; this appealed to the workers because business had been slowed by the war. The company wanted to increase prices, but prices were regulated by the government's Wage and Price Board which did not allow price increases unless a company had union representation.[1] The idea of a union was foreign to the

Grays, but they had no other alternative. By 1943 the employees voted overwhelmingly to have the AFL, a union well in place at the neighboring pulp mill, represent them. Subsequent catalogs reflected the price increases necessary to keep the workers, and although the Grays were uncomfortable with the idea of unionized labor, the arrangement seemed to be working out.

As soldiers returned home and lives returned to normal after the war, sales began to pick up. Business was increased through the late '40s and more employees were hired, including many from Chapman Machine, a long-time Old Town business that had laid off workers when its defense contracts ran out. As their numbers increased within the OLD TOWN workforce, they voted for new union representation—the AFL's International Machinists Union, the same union that had represented them at Chapman. This union was a strong one, and through the early '50s, more and more grumbling was heard about pay increases. The Grays were striving for a market share in a very competitive market and felt wage increases would be detrimental to their continued growth. In March 1954 the workers voted to strike. Picket stations were manned at each entrance of the plant, but firemen and maintenance workers were allowed to cross the lines to handle necessary work within the plant.[2] Several workers who had worked for the Grays for years, many through the Depression and the war years, felt they had been treated well. Rather than strike and join the pickets, they sought work elsewhere. The strike did not have a debilitating effect on the factory, which had built up a good inventory in the winter months prior to the strike. With the support of a local

Generations have been employed throughout the years at OLD TOWN CANOE. Tom Goslin, a superb paddlemaker for over thirty years, passed on the art to his son, Steve, who has worked for the company for twenty-two years and is now head of the wood department.
Old Town Canoe Company

trucking company, the Grays and a skeletal crew shipped orders as negotiations continued into the spring. Workers refused to accept a contract that had the same top hourly wage of $1.08 and asked for an increase to $1.25.[3] The strike lasted for sixty-three days with little local support.[4] Finally everyone

I've enjoyed using OLD TOWN canoes for over sixty years. My training began at age six, in the mid-1930s, at a YMCA camp near Penn Yan, New York, where I learned paddling and safety using OLD TOWN canoes. In 1944, when I was a teenager, my parents bought a 16-foot OTCA that we used in Canandaigua Lake for almost thirty years. It became mine in the early '70s when my mother passed away. I was living in Vermont then, and my family enjoyed many trips throughout New England with our canoe. One exciting trip was in Maine, starting at the outlet of Mattagamon Lake, through Chamberlain Lake, up the Allagash River to Allagash Lake, and back.

After the canoe had been used for almost forty years, it needed to be restored and recanvassed. I joined the WCHA, where I learned the many things necessary to restore it. I did have some challenges, however, because the canoe had been built during World War II. It had steel fastenings, because copper and brass were unavailable at that time. But the restoration was a success.

My wife, daughter, and I now own and love three other OLD TOWN canoes: 1920 and 1924 Yankee models and a 1924 Ideal model. These old wooden canoes hold some of my fondest memories.

—*Dick Haupt, Peru, NY*

I was born and brought up in Bangor, so I was well acquainted with the OLD TOWN CANOE COMPANY, starting with a canoe we had back in the '30s when my family had a cottage on Lucerne Lake.

After spending 1942–43 in the South Pacific, where I found the islands not at all like I had read about in geography classes or in the National Geographic, *I was assigned to a vessel*

going to England. We arrived in the spring of 1944, knowing we had a role in the upcoming invasion. To make a long story short, we went in, as assigned, to Omaha Beach at 10:00 A.M. on June 6, and through a miscalculation hung up off the beach on sunken LCT—for five days. Needless to say, it was a stressful period.

We returned to England on June 11, dropping off our casualties at Plymouth, and some of us got overnight leave. The gunnery officer and I decided we wanted to get as far away as possible from the ship and took a cab to an inn in the country, miles from Plymouth and the war. After a pleasant dinner, we inquired about local diversions and the innkeeper directed us to a pub in town, about a mile or so up the river. He pointed the way up a path along the river, or, he suggested, we could use one of his boats. We could see several punts tied up to the float, a dory, and a canoe. We selected a punt, but I strolled over to the canoe out of curiosity, since I had never seen an English canoe.

And there, wonder of wonders, was an OLD TOWN canoe! There is no way to adequately describe my emotions at that moment. There I was, thousands of miles from home, just hours back from the harrowing experiences of the past few days, and there was a reminder of home, right in front of me. Needless to say, we took the canoe, visited the pub, and had a fine evening. It was a memorable occasion, one I will treasure for all time.

Even now, when I see an OLD TOWN canoe, the whole thing comes back in a rush—it was wonderful!

—*Ed Hooper, Bangor, ME*

As a sailor during World War II, Allan Small and I served aboard the General H. W. Butner, *a troop transport. During a stop in Townsville, Australia, we visited a shop and the clerk asked where we came from.*

"New England," we answered.

"Where?" he asked.

"Maine," we said.

"Where in Maine?" he inquired.

"Old Town," I answered.

"I have an OLD TOWN canoe at my cottage here in Townsville," he responded proudly. "What a small world!"

—*Clair Shirley, Gardner, MA*

realized that it was a no-win situation, and the strike ended with both groups agreeing to the old contract and a modified union shop.[5] The contract now stipulated that union membership was not a requirement of employment, and many of the workers returned to work disavowing the union. Pleased at their allegiance, Sam Gray gave them a wage increase equal to others that prevailed in town.[6]

Throughout the post-war years, OLD TOWN CANOE COMPANY added new models to the line: sailboats, cartoppable sport boats, harbor boats, and rowing skiffs. More emphasis was placed on boats that were faster, sleeker, and allowed more horsepower. In 1949 an all-wood outboard was introduced. The public's interest was not on manual propulsion but on speed. OLD TOWN was also confronted with a new technology

developed during the war: riveted aluminum construction.

The Grumman Corporation had made its mark producing airplanes during the war. Following the war, Grumman diversified and adapted the aircraft-building techniques to the manufacture of canoes. Although this material was strange to traditional canoeists, it was not foreign to thousands who had served in the war or to those involved in the production of airplanes. Although more expensive, aluminum offered a canoe that was lighter and tougher than its wood-canvas counterpart. The 17-foot Grumman resembled an OLD TOWN OTCA but differed, as well. In wood-canvas construction, the thwart behind the front seat was necessary to maintain the shape of the canoe, and the seats that hung from the rails added no further support. In the aluminum model, the thwart behind the seat was unnecessary because the seat itself provided enough structural strength.

The Grumman Corporation approached Sam Gray in hopes of making a deal, but Gray felt aluminum was noisy, ugly, and susceptible to temperature extremes and rock drag. Although he was certainly entitled to his opinion, he soon found that not everyone agreed, as OLD TOWN's canoe sales sank to a new low of as few as 200 canoes selling annually in the early '60s, while Grumman produced nearly 20,000.[7] One OLD TOWN salesman reportedly said that there was a tin line running from Buffalo to Pittsburgh, and it was futile for him to cross it because west of that line, aluminum was king.

Responding to market trends, Deane felt the time was right to expand the motorboat line. Competing in the powerboat business was certainly a risk, since OLD TOWN was way behind the competition like Chris-Craft and Thompson. But boats were already 80 percent of OLD TOWN's sales, even though orders for the Sea model, which had been such a strong seller through the '20s, had dropped enough for it to be discontinued from the line. Now OLD TOWN focused on its lapstrake boats, which were nothing new to the company; an 18-foot square stern and a 17½-footer

OLD TOWN *Lapstrake Boats in 14-, 16-, 18-, and 20-foot lengths were constructed of ½-inch solid red cedar planking. Ribs were ash or oak. The gunwales were mahogany, and mahogany plywood was used for the decks. Boats were fully equipped with a ventilated, mahogany-framed windshield, complete steering, floorboards, cleats, chocks, fender hooks, lights, a flag holder, and an* OLD TOWN *flag. Curtains, storage boot, lockers, and more were some of the available accessories to complement this fine craft.*
Gray Collections

I had no idea that OLD TOWN *made a boat like the one we purchased last year: a wonderfully wide, deep, light, and seaworthy inboard/outboard! They simply called them "The Lapstrakes," and made them in 19-, 20-, and 21-foot lengths. Ours is 19 feet long, 8 feet wide, and about 44 inches deep, with one of the first MerCruiser sterndrives. The hull was built in 1963, and we're only the third owners. I restore canoes and guideboats, hate noise, and like quiet sports—but I must say we're in love with this old thing!*
—*Mason Smith, Long Lake, NY*

for rowing had been offered in the catalog as early as 1931. These models were designed by B. N. Morris, who had a thriving canoe and boat manufacturing business in Veazie until 1920, when his factory was destroyed in a fire set by an arsonist. He never recovered the loss of his own factory, but went to work at OLD TOWN and built there for a few years into the 1930s. His designs were thought to be some of the best, and his workmanship unparalleled. His association with OLD TOWN was brief (Morris died in 1940 at the age of 74), but his techniques were used for years to come.

But frustration ran high, with Deane and Jim looking for ways to increase the OLD TOWN market share, and Braley disappointed with the lack of orders. The strain of operating the business took its toll on the brothers, and Braley sought to distance himself by focusing on his accounts on the road, visiting dealers attempting to expand the distribution of OLD TOWN products. He would arrive in a new city and check the phone books for marine supply stores or hardware stores, potential outlets for OLD TOWN's line. He signed up new dealers wherever he could and continued to call on existing accounts.

The bleak days were only worsened when, on March 21, 1961, Sam Gray died unexpectedly of a heart attack. Although he was still the company's figurehead, his duties had been turned over to his sons

Denver to Old Town

Penobscot Indians escorting Ed Vestal, Bengt Soderstrom, Earl Rickers, and Gerald Hewey in 16-foot OLD TOWN Guide models to Old Town, Maine, November 11, to end their trip started May 1, 1957, at Denver, Colorado. Canoers paddled and portaged nearly 5,000 miles. On South Platte, low water forced them to push more than ride. Barbed wire fences, dams, and floods were problems. The Missouri was better. The Mississippi meant going upstream—31 days, 616 miles against 12 days, 611 miles on the Missouri. The St. Croix, Namekagon, and White rivers carried them to Lake Superior, with rapids, shallows, swamps, and log jams. Parts had not been traveled in 126 years. Georgian Bay brought gales, fast sailing, and waves 5 to 10 feet high, as did Lake Nipissing. Mattawa rapids proved exciting. Ottawa and the St.

Lawrence proved more like lake travel. The Chaudiere and Liniere to Portage Lake proved most difficult, with many portages. From Penobscot Lake, the Penobscot River led to Old Town with few portages. There was snow occasionally and ice in the puddles and dead-waters. "Canoes of any other type of construction would never have been able to stand the punishment given these OLD TOWN canoes," the canoers said on arrival.

Long challenges are not new to the OLD TOWN record books. In 1919, Lieutenant Good left Chicago, Illinois, for a twenty-month adventure in his OLD TOWN canoe. Lt. Good journeyed down the Mississippi River, across the Gulf of Mexico, around the peninsula of Florida, and up the Atlantic coast to New York, where he disembarked at the Knickerbocker Canoe Club.

Writeup and photo courtesy of Old Town Canoe Company.

Jim Cunningham takes his family for a test drive in OLD TOWN's new Lapstrake 2400 with inboard power. This beauty was unveiled in the catalog in 1965.
James Cunningham

Braley and Deane, along with Jim Cunningham. With his incredible business sense and his years of experience, Sam had remained an enormous asset to the company and the town. Even on the day of his death, he was returning from a board of directors meeting where his expertise had been needed to help reorganize the Merrill Bank. His civic work spanned decades. He had served as a member of the school board for fifty-five years, acting as its chairman on many occasions, worked as a town alderman, was

On the Road

The Gray brothers handled the bulk of the sales orders themselves. Most OLD TOWN dealers were located east of the Mississippi River and occasionally Braley flew to service the long-distance accounts in Georgia, Michigan, and Colorado. Deane handled New York by driving his sportscar to various destinations. Although a ride to upstate New York might seem grueling, Deane looked forward to getting on the Cherry Valley Turnpike, US Route 20, and listening to the hum of the engine as he cruised to Buffalo. He'd call on his dealers and then make the trip back home. Driving his favorite vehicle was a pleasure, but with the ever-increasing demands of running the company, Deane realized he needed to hire another sales representative to replace himself.

John Whitney had grown up around water and canoes. As a child he spent his summers at Camp Winona on Moose Pond in Bridgeton, Maine, usually paddling OLD TOWNs. It was no surprise that he would make a career selling marine products. As a salesman for Danforth Anchors, Whitney visited OLD TOWN often, and when he stopped by in December

1963, Deane intimated that his New York representative wasn't working out and could no longer handle the job. Deane complained further that the representative was "unreliable" and "couldn't find the time to take care of business."[8] Without asking many questions about his predecessor (who, of course, was Deane himself), and with complete confidence that he could do the job, John gladly accepted the opportunity to go to work for OLD TOWN. It wasn't until the first stop of his new route that he realized Deane's joke!

After four years, Deane extended Whitney's duties to include New England. At that time, John felt the two territories were too much for one man to handle and suggested that Deane contact John's long-time friend, Jack Gillespie, to handle New York. John Whitney's association with OLD TOWN lasted until his retirement in 1996, and Gillespie continues to represent OLD TOWN in the New York and mid-Atlantic states, which he has been doing now for more than thirty years. The two men can be seen in recent OLD TOWN catalogs paddling a red wood-canvas canoe, always their favorite.

One crisp fall day in 1951, two brothers went driving down the Paw Paw River hunting ducks in their 18-foot 1921 OLD TOWN canoe. The younger brother was in the stern, while the older brother sat in the bow, shotgun at the ready. Suddenly a duck flew up! The bowman followed the duck's flight until his twelve-gauge was pointing 90 degrees to the long axis of the canoe, at which time he fired (missing the duck). In accordance with Newton's Third Law, the canoe promptly tilted the other way, dumping the unprepared sternman, his paddle, and his sixteen-gauge double-barreled shotgun into the river while the OTCA righted itself, contained the bowman, and didn't take on a drop of water! To the older brother's amazement, the younger brother demonstrated that it is possible to swim to shore even while weighed down with full hip boots. Maybe the frigid water drove him. The younger brother and the paddle were recovered, but, alas, the sixteen-gauge remained forever at the bottom of the river. —Jim Woodruff, Lansing, MI

Nearly all people have dreams, but only a few put them into practice. As an adult with dreams of my own, I respect my father for living out his own dream—something I didn't understand in 1965 when I was eleven years old.

My father left his humdrum life as a copy editor and pursued his own obsession, which was to cross the moody Atlantic Ocean with his diminutive but sturdy OLD TOWN sailboat, christened Tinkerbelle. [See pp. 92–93.] As a young lad, I couldn't appreciate the danger of the proposed feat. We were sworn to secrecy until he was safely out to sea; our announcement of his escapade was soon echoed in every newspaper and television station throughout most of the western world. Nearly overnight, he was a celebrity. Naturally, this affected the rest of the family and me. It was a puzzling situation, being yanked from a complacent, low-profile existence and thrust into the media spotlight as the son of Robert Manry!

Perhaps someday, like my father, I can find the strength and courage to live out my own dreams.

—Doug Manry, Cleveland, OH

director of the Merrill Trust Company for twenty-seven years, and was a director of the Bangor Hydro-Electric Company, all while managing the OLD TOWN CANOE COMPANY. The town mourned its loss. Like their fathers before them, Deane, Braley, and Jim carried on, but it went without saying that the man who made OLD TOWN CANOE a giant in the industry could never be replaced.

By the early 1960s, the bills at the canoe factory were mounting and Deane struggled with the finances. Loans taken in the winter to produce new stock often had to be carried over when sales did not meet expectations and expenses soared. Producing lapstrake boats was a time-consuming operation. Each plank had to be hand cut, and it was a painstaking process cutting the bevels to fit. Strakes were not cut ahead of time; instead, two men would work on an order, one of whom was skilled at cutting the planks.[9] Each time a plank was cut, the craftsman would cut a mirror plank for the other side. It was a tedious process. If the company intended to be competitive, it had to find a way to decrease the time and the cost required to produce these boats. Jim Cunningham put all his energy into making a machine that would cut the lapstrake planking for the several models offered, but producing a machine to cut the planking bevels was a costly, frustrating venture. Despite the grueling machining involved, Cunningham was successful. Using cams, the machine tipped the head of the cutter to produce the necessary angles. With this tool the

The Little Lapstrake That Could

No book about OLD TOWN would be complete without a section devoted to Robert Manry and his 13¹/₂-foot sailboat *Tinkerbelle*. Manry, a copywriter for the *Plain Dealer* in Cleveland, was a man who spent most of his time behind his desk in a windowless office. In 1958 his love of sailing (and his limited budget) prompted him to answer an ad for a 30-year-old OLD TOWN lapstrake sailboat that needed some repair. Manry found the boat to his liking—small, trailerable, and cheap! With two large cracks on each side running the length of the boat at the waterline, over a dozen cracked ribs, and another six or more with dry rot, the boat needed serious attention, but he was up to the job.

The Manry family christened the boat *Tinkerbelle* because the name reminded them of the classic tale of *Peter Pan* and the courage to believe as seen in the tale. Manry also felt the name was apt because he was always tinkering with her.[10] Over the years he added new sails, an extended mast, a new daggerboard keel, and even a small cabin. As *Tinkerbelle's* appearance improved, so did Manry's sailing skills, and they began taking increasingly longer trips together.

In early 1964 a friend of Manry's who owned a 25-foot cruising sloop proposed that the two men sail together to England. Manry was ecstatic and asked for a leave of absence from work for the summer of 1965. But his friend ended up canceling the trip, feeling it required too much of a time commitment. Manry was crushed but then thought, "Why not make the trip in *Tinkerbelle*?" Without telling anyone except his immediate family about his change in plans, he continued with preparations. The evening before his departure, he mailed letters to his executive editor and to friends at the newspaper revealing the truth. On the morning of June 1, 1965, he set sail from Falmouth, Massachusetts, for its sister city in England.

Although Manry was well prepared with supplies and the necessary skills, nothing could have readied him for what lay ahead. During the endless hours of sailing, Manry pondered every question of life. He later reflected that the voyage was a kind of microcosm of life—a life within a life—that gave a sailor "an opportunity to compensate for the blemishes, failures and disasters of his life ashore."[11] His trip included a near escape from a collision with a large ocean liner, days without wind, days with too much wind, hours of being off course, and several capsizings. And then there was the excruciating loneliness and even haunting hallucinations. Despite it all he endured. While Manry grappled with his personal struggle, the world was keeping tabs on him, anxious for him to succeed. There were sightings of *Tinkerbelle* from time to time; an ocean liner relayed some of his

mail, and another ship's captain actually came aboard *Tinkerbelle* and shared what was a virtual feast.

Manry never expected the outpouring of support that he received. His family had been flown to England by the *Plain Dealer* to greet him. Television newsman Bill Jorgensen came out to sea to interview him. Aircraft spotter planes circled above as he approached the English shore. Finally, on August 17, boats of all sizes sailed out of the Falmouth Harbor to meet him. As he approached the docks of Custom House Quay, "people were everywhere: standing along the shore, perched on window ledges, leaning out of doorways, crowded onto jetties, thronging the streets, clinging to trees and cramming the inner harbor in boats of every size and description."[12] Fifty thousand of them had come to welcome Manry and *Tinkerbelle* to their shores. Manry had fulfilled his lifelong dream, and *Tinkerbelle*, a 13 1/2-foot OLD TOWN sailboat, had made it possible.

builders could streamline the process and cut the cost of production.

But their triumph was short lived. No sooner was the machine capable of producing a specified lapstrake design, than the public wanted something different, either wider or deeper. New models required new patterns, and the process of altering the cutting jigs was endless. New jigs were created, the machine would be altered, but all of this tooling was costly and time-consuming. It was a machinist's recurring bad dream. OLD TOWN couldn't keep up with the whims of the general public nor with competition from manufacturers producing boats made from plywood.

[1] Braley Gray Interview.
[2] "Strikers Picket Old Town Canoe Company Plant," *Bangor Daily News*, March 23, 1954.
[3] "Pickets Are Stationed At Canoe Plant," *Penobscot Times*, March 25, 1954, p. 1.
[4] "Old Town Canoe Strike At End," *Bangor Daily News*, May 22-23, 1954, p. 2.
[5] *Ibid.*
[6] *Ibid.*
[7] Tim McCormack and Alan Hirsch, "A 'Discovery' Revitalized Old Town Canoe Company," News Release, Phillips 66 Company, August 10, 1987, p. 2.
[8] John Whitney Interview, February 1994.
[9] Jim Cunningham Interview.
[10] Robert Manry, *Tinkerbelle* (New York: Harper Row Publishers, 1966), p. 17.
[11] *Ibid.*, p. 67.
[12] *Ibid.*, p. 226.

OLD TOWN had long kept current by turning out new models, but by the early '60s it seemed that the time had come to consider new synthetics—and the new miracle material was fiberglass. Still reluctant to abandon wood-canvas, the material that made the company great, Deane Gray watched the industry's early use of fiberglass closely. A sign on Deane's desk said, "If God wanted fiberglass boats, he would have made fiberglass trees!" Not everyone shared this sentiment.

His closest neighbor did not waste any time before using fiberglass. The White Canoe Company, long-time builders of wood-canvas canoes and boats, was employing the new technology through the abilities of Walter King, better known to his friends as "Bub." King had worked for White for a number of years, and after the war, he and his brother-in-law Pat Farnsworth, a local businessman, purchased the business from E. M. White. Together they were to enter a new era of production.

King's building expertise was remarkable. His achievements in fiberglass boat building and design were unparalleled, and he helped to restore White to the level of prestige the company had once held. But even though sales were improving, the increased appeal of aluminum brought competition, and, of course, interest in faster, bigger boats challenged the design department. Farnsworth decided the boat business was not for him and sold his partial interest in the White Canoe Company to Herb Sargent, another local entrepreneur who put his son-in-law John Gowen in charge of overseeing his interests.

By 1964 a slowdown in the industry was taking its

12

Old Town Tries Fiberglass

toll. White had run into financial trouble making the transition from wood production to fiberglass. King felt pressured to carry on the White tradition of building fine boats, but with his back to the wall, he had no other choice but to abandon the struggling company.

It was a very difficult decision for him. His family was comfortable in Old Town, the children attended local schools, and he knew everyone in town. He had repeatedly turned down offers from Chris-Craft and Bertram, who had recognized his talent. Instead, he liked it simple, and he liked having the reins. Deane Gray promised him just that, so King sold his interests in the White Canoe Company to Sargent and went to work for OLD TOWN CANOE.

With King on board, the Grays had a new opportunity, one they felt they badly needed. OLD TOWN was financially strapped. The loan interest was mounting, and it seemed harder and harder to pay the bills as raw material prices grew but the demand for new canoes and boats did not. Adding to the company's woes was a shortage of Sitka spruce brought on by Japanese competition for the wood, coupled with Alaskan earthquakes. King, with his experience in man-made materials, held the answer, and he began

designing a new canoe—the first OLD TOWN made of fiberglass.

King's design, featuring a molded keel and a deck cover with molded seats, was revolutionary. His brainchild was a one-piece molded top that included the rails, thwarts, and seats.[1] The canoes were assembled from only two molded fiberglass parts, a one-piece hull with a molded keel and the one-piece top. Polyurethane foam, expanded under the gunwales and decks, gave the canoes upright flotation three times their hull weight. It was a bit impractical—water didn't drain from the canoe easily—but despite some minor imperfections, the design successfully utilized many of the design benefits of fiberglass-reinforced plastics. King's new 16-foot and 18-foot fiberglass models were award winners in no time. The lines were long and sleek. It was the unanimous choice of four notable Chicago industrial designers, who awarded OLD TOWN's fiberglass canoe the Grand Prize of the Society of Plastics Industries, Reinforced Plastic Division, for 1966. The canoe was selected for excellence in originality, application, and design. It beat out such prestigious industry notables as the futuristically designed Studebaker Avanti and a state-of-the-art missile nose cone. For the OLD TOWN CANOE COMPANY, it was a very celebrated win and the turning point of a new era. Deane Gray accepted the Steuben-designed Corning glass trophy proudly.

Deane finally had the edge he needed, but the canoe line was not to be King's priority. Deane hoped that Walter could develop a fiberglass boat as well. King wasted no time bringing his experience at White

Deane Gray at the helm of the company's motorboat jewel, the Atlantis 2560, constructed of fiberglass. Gray Collections

into play. With the help of Jim Cunningham, who was largely responsible for all the OLD TOWN lapstrake motor boats to date, the two men went about designing a new powerboat. They scrutinized the competition at the yearly Madison Square Garden Show and spent long hours discussing what they wanted to see in a large powerboat. Together they built a wooden prototype and took the boat to the ocean at Winterport, Maine, to test its on-the-water performance. The trials were disappointing. The V-design of the hull was not deep enough, and the boat pounded rather than sliced the water. When they returned to the factory, they took an ax to the hull, modifying it significantly. After days of reconstruction, they tried the water tests again. This time the boat responded in the way they had hoped. The final product was still weeks away from completion as they readied the form and plug, but at last the fiberglass Atlantis—24 feet of "agile

Bathing beauties pose with an OLD TOWN canoe in the 1960s.
Old Town Canoe Company

white magic"—was finally unveiled. It had a deep-V fiberglass hull that ran steady in chop or heavy seas, just as its designers had planned. Billed as a pleasure craft with lots of room, the boat featured a plush cabin fully carpeted from floor to ceiling, and it was further equipped with two large cushioned bunks. For those who wished to use the boat for work, large color-keyed storage compartments ran along both sides with more below to stow fishing or skin-diving gear easily. Not to be outdone by the competition, the Atlantis was powered by single or double inboard-outboard engines. With the throttle wide open on a 210-hp Evinrude or Johnson engine, it reached 38 mph. The Atlantis was a significant achievement for OLD TOWN;

the company hoped the public would agree and waited for the orders to come in.

Other changes were happening rapidly at OLD TOWN, and a combination of circumstances caused Braley to decide to sell his share in the family business. In 1967, through arrangements made by the bank, he sold his stock back to his siblings and received a contract that allowed him to work for as long as he wanted, continuing the sales and road shows for which he was so well suited. For five years the compromise worked successfully. Then Braley retired.

[1] Jim Cunningham Interview.

When I started planning a trip from Lake Athabasca down the Dubawnt and Thelen Rivers to Chesterfield Inlet on Hudson's Bay in 1966, I asked Deane Gray whether OLD TOWN would be willing to modify one of its designs to accommodate the special needs of our trip. He was intrigued with the idea and agreed.

What we had in mind was to deepen the 18-foot Guide Special by 2 inches and to adjust the height and location of the seats to meet our packing needs. The deepening was the most important, because our route would take us across some rather large and unprotected waters: we had to paddle the length of Lake Athabasca, we were going to circumnavigate (almost) Dubawnt Lake, and we planned to paddle at least 200 miles on the open waters of Hudson's Bay. In addition, our loads were going to be substantial because we had to paddle two months without reprovisioning on the 900 miles from Stony Rapids north to Baker Lake. Deane Gray offered to modify the Guide's standard dimensions and loan us the canoes for the trip.

OLD TOWN manufactured two modified 18-foot Guide canoes (#175696 in green, and #175699 in red) starting in early February 1966 and completed the final varnishing on March 14. Both canoes had outside stems as well as "shoe keels" to provide some modest directional stability on the big waters. The canoes were shipped by rail on April 7 from Old Town to Waterways, Alberta. They were then taken by barge to Uranium City, Saskatchewan, where we picked up the canoes on July 10, uncrated them, bolted on the yokes, and set off on our trip the next day.

The canoes functioned extremely well on our ten-week trip. We never shipped water, even in large rapids or storms or the confused-wind tidal waters of Chesterfield Inlet. They were competent, quiet, and fast on the flat. When we completed the journey at Chesterfield Inlet in late August, we left the canoes at the Hudson's Bay Company post there for shipment by ocean freighter north out of Hudson's Bay and down to Maine. That was the last we saw or heard of them for over thirty years.

This year, after making Benson Gray's acquaintance at the WCHA bulletin board, I learned that he was Deane Gray's son and he agreed to do some searching for information about the Dubawnt canoes. What he learned was very interesting. The boats had been built on the OTCA forms, not the Guide forms, and he thought that our modifications may have inspired OLD TOWN's later design of the Tripper, which, oddly enough, was originally designed on the OTCA form as well.

Using Benson's data from OLD TOWN's sales records, I learned that the canoes were returned to OLD TOWN in late 1966 from Chesterfield Inlet and were on display at the factory for about eighteen months. They were then refurbished to some extent and sold, one to a University of Maine student and the other to a paddling enthusiast from New Hampshire. I have been able to trace the New Hampshire canoe (the original green one) through four sets of owners and twenty-two years from 1968–1990, all the way out to Livingston, Montana. It was last located at a ranch that burned down in 1990, and I've been unable to track it further. I'm now going to concentrate on the red canoe that was sold to the University of Maine student. I'd love to see how one or both of these canoes have survived the years!

—Robert B. Thum, Lafayette, CA
(Benson Gray tracked down the red canoe on the Internet; it showed up at a restorer's shop in New Hampshire. Bob Thrum promptly bought it, and it is now being restored.)

13

The Genius of Lew Gilman

With Walter King busy with the fiberglass production of the larger boats, Deane needed support in producing the new fiberglass canoes. He turned to a local man who was already stirring things up in a big way, the self-taught, energized genius Lew Gilman.

Lew was born in Old Town in 1929 and grew up there taking advantage of the fishing, paddling, and hunting the area offered. For most of the kids in town, the OLD TOWN factory loomed invitingly, like a giant playground. It was a great place to explore, play hide and seek, and even pick up some extra scrap that could keep any kid occupied for a while. When winter dragged on relentlessly, the boiler room provided a penetrating heat that was a respite from the cold, uninsulated walls of home. There were no locks on the doors, no signs that said, "Keep Out." On the contrary, anyone could walk into the leviathan-like plant—and they did, young and old alike. Workers would often find themselves shadowed by curious onlookers. As long as visitors didn't get in the way of the workers, they were tolerated.

For young Lew the factory was always an adventure. He loved to wander about and often found some scrap to tinker with, making one contraption after another. Johnny Moore, who had a machine shop in town, recognized the young man's ability and encouraged him by giving him a job when he was fourteen. Lew often delivered machinery or parts made at Moore's to the canoe factory, and this association kept him a familiar face with the workers in the plant.

The Korean War interrupted life for many of the townspeople. Lew was no different. After serving his one and a half years in the Air Force, he anxiously returned home to resume the life he'd left behind. His military contacts and increased skills secured him a position as a civilian welder for the Air National Guard. Not long after, he became the owner of an automotive repair shop. Cars had always been a hobby, and Lew's theory was that if they could build them, he could fix them. The automotive business also gave Lew more practice working in fiberglass. He could mold fenders and do body repairwork using it. Fiberglass seemed limitless in its applications.

Lew's love of paddling had been lifelong, and after tedious, repeated patching of his wood-and-canvas canoes, he began to think about manufacturing canoes in fiberglass. He discussed the possibilities with his long-time paddling friend Paul Rivers, and they decided they would give the process a try, setting up shop in a former chicken coop, and calling the business Rivers and Gilman of Hampden, Maine. One of their first fiberglass models was an 11-foot canoe, which was successful as a product but not a great moneymaker. The canoe was made in a birchbark finish that didn't show dirt or scratches and was attractive to outdoorsmen. Despite slow sales, the two partners persevered.

The target of Rivers and Gilman's sales was the novice canoeist, and they were content with their small part of the market. Lew was the chief designer. His favorite was the 17-foot Princess model, developed from an old Morris form for which he owned the rights. It was a pretty canoe, refined in its lines like an old classic but not very versatile. Although Rivers and Gilman specialized in fiberglass construction, Lew knew the material could not live up to the rigors of Maine and the canoeing demands to which he was accustomed. With one paddling miscue on the rocky northern rivers of Maine, the Princess could be quickly reduced to colored fragments traveling downstream.

Meanwhile, at OLD TOWN, the Grays and Jim Cunningham were watching Gilman closely. On more than one occasion, Cunningham asked Lew to come work for OLD TOWN, but he refused. Despite OLD TOWN's clout in the industry, Lew was not interested in working there using the current materials, and despite his successes with fiberglass, he felt there had to be something better out there.

Lew Gilman has been a key figure in revolutionizing canoe and boat construction at OLD TOWN CANOE.
Old Town Canoe Company

At a Chicago boat show in 1964, the Thompson Boat Company of Peshtigo, Wisconsin, and Cortland, New York, unveiled a canoe made of Royalex,[1] a virtually indestructible vulcanized composite of Acrylonitrile-Butadiene-Styrene (ABS), a material developed by the Uniroyal Company of Indiana.[2] Lew examined it closely. He recognized the design to be a modified 16-foot OLD TOWN but felt that the canoe presented in this material was too flexible. It had a 4-inch deflection just sitting still, never mind how it would flex in the water! He ordered one for closer examination on his return to Maine, and decided that although Royalex was a potentially ideal material, being extremely tough, it needed more work if it was going to be used for a functional canoe.

Representing Rivers and Gilman, Lew tried to talk Uniroyal into making canoes for his company, but Rivers and Gilman was too small to interest Uniroyal. Instead, Uniroyal attempted to convince OLD TOWN to take a look at the materials, but discussions with Deane Gray went nowhere. Lew Gilman didn't give up. He approached Uniroyal with design changes that he felt would greatly improve their canoes. He talked them into removing the keel, which had actually weakened the bottom by compressing the supporting underlayers. He also convinced them to add one more layer to the floor to reduce its flexing.

Dean Hawley of Uniroyal's sales department was impressed enough with Lew's ideas and perseverance to make Rivers and Gilman six canoes in a trial run.[3] The thermoform process was used to produce white hulls; black was added later to produce a birchbark-like pattern. (They were also available in red, blue, yellow, and green.) These new canoes were marketed under the Indian brand name by Rivers and Gilman and several are still around today, a tribute to their durability. The orders didn't help Uniroyal's bottom line, but the product proved itself marketable.

Despite the success of the new material and increased sales that now reached approximately 1,600 canoes in 1968, all was not well at Rivers and Gilman. The partnership was on the rocks, and to make matters worse, the company suffered a fire in October 1968. Everything was lost. When Gilman learned that the company was only partially insured, he left the partnership, even though it meant taking a substantial loss.[4]

In December 1968 Lew finally agreed to go to work for OLD TOWN. Deane Gray hoped that with the addition of Gilman, Walter King could concentrate on building boats. Lew felt the company might be after his Royalex experience, but no matter—he would be doing what he liked best. Building boats with Walter, a man he greatly admired, would be a tremendous opportunity, so he gladly accepted and hoped 1969 would be a brighter year.

The Fastest Tack Spitter in the Industry

Although the focus had changed to fiberglass, OLD TOWN still built wood-and-canvas canoes and for many years the highlight of an OLD TOWN factory tour was visiting the renowned craftsman Clyde Hinckley. Clyde was a planker and had the reputation of being the "fastest tack spitter in the industry." It often took four men to keep up with his production capabilities. Clyde would store a number of tacks in the left side of his mouth and then move them with his tongue so he could take them out point first. Next, he would push the tack into the boat with his fingers, and hit it quickly with his hammer. When he had an audience, he really put on a show, spitting out tacks faster than the eye could see. On occasion, he would make a mistake and pound one in crookedly, but never let on to his admiring public. Instead, he would spend time later digging out the errant tack. On several occasions, he swallowed a tack but never admitted it. His fellow workers knew he was distressed when he ate plain white bread. The "old-timers" believed that bread helped the stomach cope with the digestion of the metal tack.[5]

Because the men were paid by the piece, Clyde prided himself in doing three canoes a day, equaling about 4,500 tacks! Clyde was also known to pick up extra money by working at night when the factory was closed. He'd find an open window, sneak in, and go to work. The other men would arrive in the morning and find that they were already well behind.

Clyde's ability was so remarkable that whenever a new tool salesman came in with a power tacker, Deane would send him to Clyde. After seeing what Clyde could do without the use of a power tool, the salesman quickly gave up and departed.

[1] Royalex is now a product of the Royalite Thermoplastics Division of the Uniroyal Technology Corporation of Mishawaka, Indiana. Royalex is a registered trademark.
[2] Lew Gilman Interview.
[3] *Ibid.*
[4] *Ibid.*
[5] George Cook Interview.

14

A New Game Plan, a New Team

Instead of canoes, Gilman found himself building kayaks at OLD TOWN under the guidance of Bart Hauthaway. This was not the first time the company had offered a kayak; in 1940 a catalog insert had introduced a 10-foot wood-canvas model. By the late '60s, European involvement in whitewater sport was already strong, and Deane Gray felt it wouldn't be long before Americans would "catch the bug." He hoped to get in on the ground floor to capitalize on the growing interest in whitewater boats and hired Bart Hauthaway to introduce a new breed of kayak to the OLD TOWN line. These kayaks would be constructed of fiberglass with semi-flexible resin for more durability. Hauthaway would act as a design consultant, and OLD TOWN would set up production. In their gentlemen's agreement, Hauthaway could still produce and sell custom kayaks from his home.

Bart Hauthaway's boats already had a strong reputation, not only in his native New England but throughout the paddling community. Bart protested that he never intended to build commercially, but like so many great builders before him, found that each time he built a new boat, someone would buy it. At a time when few companies were producing kayaks, his models excelled and were thought to be some of the most innovative. He was also a successful racer.

Whitewater World Championships are held every other year. Bart was a member of the United States team in 1965 and was back as the designer of the team's boats and its co-coach in 1969. For the athletes the Whitewater World Championships were a chance to showcase their increasing skills. For OLD TOWN it provided an opportunity to capitalize on the new en-

thusiasm for whitewater sports by supplying the team with kayaks. This was the first time that a canoe/kayak team had a corporate sponsor, and Deane was proud to support the U.S. team's efforts.[1]

Despite its lack of international experience, the United States team had made its best showing to date at the '67 Championships, where it ranked 23rd out of 42 in K-1,[2] 9th out of 19 in K-1W, and 13th out of 26 in C-1 slalom. With further competitive experience and Hauthaway's designs, there was a vast improvement at the '69 championships on the Isere River at Bourg Saint Maurice, France. The Americans surprised everyone by placing 10th in the K-1, and winning two bronze medals in the C-2 mixed wildwater and slalom team events. Team members displayed new confidence in their OLD TOWN boats and in their ability to compete with the world's best.

Too late for inclusion in the 1970 catalog, OLD TOWN issued a separate catalog for its new line of kayaks and canoes called "New, Now, Wet & Wild." The tri-fold layout provided readers with six pages of

photographs, from surfing at Cape Cod to the 1969 World Championships in France. Four models of kayaks were offered, three of which met competitive specifications. The Slalom, at 13 feet, $1^1/2$ inches, complied with the international rule of "not less than 4 meters" and was short and maneuverable. The Downriver and Downriver Touring kayaks at 14 feet 8 inches were longer and built for speed. Hauthaway also added an 11-foot Junior kayak to help encourage youngsters to participate. The catalog promised that by late spring of 1970, OLD TOWN would produce C-1 and C-2 slalom and wildwater canoes.

Kayaks and canoes could be ordered in any of thirteen different colors and/or combinations for hull and deck. The colors included celery, avocado, jade, turquoise, blue, russet, red, dark green, mahogany, yellow, sandalwood, desert white, and white—a wide spectrum reflective of current trends. Deane Gray went so far as to offer a psychedelic paisley pattern. He really wanted this new line to stand apart from anything the company had ever done. All the latest in whitewater or surfing equipment was available, including a contoured boat carrier designed by Hauthaway himself.

Deane Gray's attempt to bring OLD TOWN back to the manufacturing forefront covered all the bases: wood-canvas canoes, fiberglass canoes, several models of kayaks, sport boats, and, of course, the powerboat line that included both inboard and outboard models.

Offered in 1940 through a catalog insert, this kayak is similar to OLD TOWN's current Otter model in size and deck shape. Gray Collections

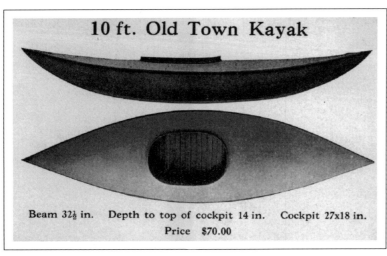

10 ft. Old Town Kayak

Beam 32½ in.   Depth to top of cockpit 14 in.   Cockpit 27x18 in.
Price $70.00

Unfortunately, just when things were looking up, Walter King suffered a debilitating stroke and had to retire. He had established a routine making the big boats and the fiberglass canoes, and the work continued without him, following his methods. Joe LaVoie, a long-time friend and co-worker from the days when they both had been at White, had followed King to OLD TOWN and now assumed many of Walter's boat-building responsibilities. LaVoie was also a well-respected mold-maker.

Lew Gilman was handling the construction of the kayaks, and with Hauthaway's input, more models were added to the line, including a Surf Kayak in 1971. This shovel-nosed kayak was easy to maneuver and rode the northeastern waters as well as a surfboard. It piqued much interest when it was seen in action, but dealers were unaccustomed to selling something so unique. They didn't know how to present it to the public.

A diverse selection of whitewater canoes was also offered in 1971—from the Wenatchee, a solo closed model, to the Potomac and Berrigan, which were made to carry two canoeists; and finally the Ojibway, which was an open craft in both 13-foot and 16-foot models. The canoes were constructed reflecting King's ideas and were consistent with other canoes made by the company at that time. An end-grain balsa core was sandwiched between the fiberglass layers to create a rigid, strong bottom. These closed canoes were also equipped with built-in bucket seats, toe blocks, knee-pads, and quick-release knee straps. OLD TOWN was making a concerted effort to introduce new products to the paddling public, but the market was still small.

Although the line was received well by whitewater paddlers, the mass frenzy that Gray had hoped for hadn't happened yet.

In the meantime, Hauthaway saw other markets that had not been tapped. He designed a fiberglass Rushton model, 10$^1/_2$ feet and a mere 18 pounds. This featherweight canoe was patterned on the "Wee Lassie," an extremely light Nessmuk model sport canoe built in 1893 by J. Henry Rushton of Canton, New York, the premier builder of the time. This OLD TOWN model has taken its place in boating history and is now displayed side by side with Rushton's original "Wee Lassie" at the Adirondack Museum in Blue Mountain Lake, New York. The display celebrates Rushton's originality of design and applauds the creativeness of builders like Hauthaway who utilize new materials.

Boats that were light and manageable like Rushton's made it possible to get away to the serenity of the wilds. An avid hunter, Hauthaway also designed the Widgeon, for those who shared his love of waterfowling. This new 9-foot duckboat was a one-man gunning skiff for marsh, river, or open water shooting. In a marsh green color, it was nearly invisible in the grass, or it could be purchased in white for winter shooting. In spite of these successes, a rift began to grow between Gray and Hauthaway. OLD TOWN dealers had begun to complain about Hauthaway's custom kayak building and retail sales, and Deane pressured Hauthaway to limit them. By 1974, despite the large display of his designs in the catalog, there was no mention of Bart Hauthaway as the designer. He was no longer used as a consultant, and the gentle-

men's agreement ended up in an out of court settlement.[3]

Lew Gilman had gained more experience making kayaks, but was busy with a variety of other jobs as well. His first assignment had been to invent a new deck configuration for the fiberglass canoes. The full top that had won honors in 1966 was cumbersome and time-consuming to install, so Gilman worked on rails that would provide enough area to attach the seat and thwart. Jobs like this were a satisfying challenge, but Lew never abandoned his hopes of working with Royalex and felt he could improve upon Royalex's methods. Royalex relied upon a heavy iron frame to keep the sheet rigid as it was thermoformed. After heating the sheet, the softened Royalex was forced into a right-side-up mold. Gilman felt this technique

The Widgeon, a 9-foot gunning skiff, was designed by Bart Hauthaway for OLD TOWN. *Gray Collections*

was cumbersome and ineffective. A mold constructed right side up collects dust, dirt, and worse. On one of Lew's visits to Uniroyal, a worker lost a cigarette lighter—which reappeared molded into the bottom of a canoe.

Gilman was sure he could improve on the process and set about to create machinery in which the heated sheet of Royalex would be vacuumed into an upside-down female mold to give the sheet the shape required. The process would be quick, efficient, and less costly. Deane gave him the go-ahead for whatever equipment he needed—as long as Gilman "did not spend any money."[4] Gilman tried to comply by using auto and machine parts that were lying around. The frame was built from old railroad track. He used chain falls and chain binders to move the equipment, and developed a mold from one of Hauthaway's slalom kayaks. With the use of a shop vacuum, Gilman "popped" the first thermoform kayak, later christened the Snapper, hot out of the oven. Uniroyal's experts were skeptical, but he proved them wrong time and again as the kayaks came from the oven. Gray was impressed enough to give Lew the go-ahead to spend some money to build a more permanent machine. It wasn't long before Lew was using the process to mold canoes. His efforts resulted in the Chipewyan 16-footer, later known as the "Camper," and Chipewyan 18-footer, the first OLD TOWN canoes made of Royalex.

For decades OLD TOWN had used diamond-head brass bolts to secure thwarts and seats in wood-canvas canoes and boats. It is often the first clue indicating that a canoe or boat might be an OLD TOWN. Continuing the tradition, the new Royalex Chipewyan 16 was assembled using the traditional diamond-head brass bolts, and its rivets were machined in the shape of diamond heads to secure the vinyl gunwales. But the process proved tedious, and as the orders began to mount for the new canoes, the men returned to using the standard round head rivets instead.

The word was soon out that there were canoes on the market that could really take abuse but weren't made of aluminum. They carried the reassuring OLD TOWN name, an assurance of quality. The two models were costly for the time—$330 for the 16-footer and $365 for the 18-foot version—but the promise of longevity made them worth every penny. Sales increased rapidly.

The Snapper kayak in Royalex had proved itself equally durable, but dealers seemed apprehensive about endorsing a sport about which they knew so little. The public found the kayak a bit too heavy to lift comfortably, and orders for the Snapper were flat.

With Hauthaway gone, Gilman's role in research and development became even more essential. Gilman felt that Royalex would be in widespread use soon, and it was important for OLD TOWN to lead the market. Just as part of their business was to attend sportsmen's shows, Lew Gilman and Deane Gray also attended local whitewater races throughout the region to get a firsthand look at how the sport was progressing. They saw a need for a high volume canoe that would be responsive. Almost the only canoe fitting that description at the time was the Chestnut Prospector, a legendary wood-and-canvas canoe from Fredericton, New Brunswick, Canada. Gilman studied it carefully, and although he favored a 20-footer made by White, he knew the longer length was impractical. With a development team that included Seth Gray, Deane's eldest son, Lew began the new model by using a 17-foot OTCA canoe as his working plug. Although his first intention was to rocker each end, he felt the banana-like shape would make the canoe good for whitewater but difficult to paddle well in calm stretches. Instead, he added a bow and stern with the more graceful lines he'd admired on some of Rushton's canoes. The prototype canoe, built in fiberglass, still had enough rocker for turns, but it was very subtle.

Production began in Royalex in 1973 and Gilman gave the prototype Tripper to his friend Walt Abbott, a whitewater-canoeing instructor at the University of Maine, for further testing. Walt challenged the

A bit ahead of its time here, surf kayaking is growing rapidly in this decade.
Gray Collections

In Royalex construction, (a) above left: the heated sheet of Royalex is placed beneath the mold; (b) above right: the mold is lowered into position, and a vacuum pump pulls the Royalex sheet up against the inside of the mold; (c) right: after the curing process, the mold releases the canoe blank; (d) next page left: the excess Royalex is trimmed away, and (e) next page right: finish work begins installing thwarts, rails, etc.
Old Town Canoe Company

I entered my fortieth year in 1985 with a mid-life vow to become less corporate and more independent. Where to start? Well, with an OLD TOWN Penobscot. This wonderful canoe accompanied me on trips to the Boundary Waters and several rivers in Minnesota and Wisconsin. It received the same care and attention that my two children received a generation earlier. I named her the Green Hornet and added decals depicting that name along with a green-hued hornet with a ferocious stinger. What fun to receive comments from other passing canoeists!

Eventually, I moved on to the joys of solo canoeing and sold the Green Hornet to a young couple who didn't have to search for a paddling partner, as I did. The decal is so distinctive that I have recognized her from afar, since then.

OLD TOWN makes great canoes. I have toured their plant several times while enroute to paddling Maine rivers. I've paddled many miles in other OLD TOWN models, but none are quite the same. Maybe the first experience is the best!

—Jim Anderl, St. Cloud, MN

canoe to every move he could imagine and came back impressed. He told Lew that it was the "best white-water canoe" he'd ever used. With Abbott's recommendation, Lew was optimistic, but he wanted other opinions as well. Gilman, along with Seth Gray, took the canoe to a local river for some testing of their own. They swamped and wrapped the canoe several times and saw that the hull came back to shape but that the gunwale had snapped. Lew remedied this by adding an aluminum insert inside the rail to strengthen it, then it was back to the river with the canoe for further tests.

A 150-yard run of Class III rapids was the proving ground. After filling the Tripper with water, they released it down through the rapids five different times, each time resulting in a similar scenario. After bouncing from rock to rock, the canoe took on water, and flipped. At one point it became wedged on the rocks, and the men had to use a long bar to pry it off. Each time, the canoe and its reinforced rail withstood the punishment.

Their final test was to actually allow the water-filled canoe to wrap around a bridge abutment. Within seconds of hitting the abutment, the canoe was wrapped around the concrete pillar by the force of the water. The pressure caused the bottom of the canoe to touch the thwart—the canoe was literally flattened and folded nearly in half. After a great deal of physical effort, the men dislodged the swamped prototype, and as it came off the bridge, the canoe straightened out and sat floating proudly in the middle of the stream. Gilman knew he had a winner.

The two men returned to the factory to boast about their accomplishment. Deane was pleased but wanted

to see for himself, grabbing his camera to record the event. They returned to the same spot to repeat the process, this time with several skeptical workers in tow. As they broached the canoe again, Deane snapped away. Again the boat was resurrected, slightly worse for wear, but ready to be paddled.

As they returned with the canoe to the factory, one of the crew dared that the canoe would not survive being thrown off the roof. Not wanting to be outdone or doubted, Lew Gilman and Seth Gray carried the canoe to the roof and proceeded to throw it off for a 35-foot drop to the ground. Deane captured the episode on film for all to see. As it sailed downward, the canoe pitched, landing headfirst. The bow buckled and folded but with little damage, and the hull snapped back into shape. The only visible signs of the adventure were a few scrapes and a cracked deck. The photos of that test were used for years in OLD TOWN catalogs and advertising cards.

The success of the Tripper is a testament to the strength of Royalex and the perseverance of Lew Gilman, a man who does not accept "It can't be done."[5] With sales of the Tripper having a very positive effect on OLD TOWN's bottom line, Lew was rewarded for his efforts with a raise and what he considered a greater reward—a compliment from Deane Gray. The Chipewyan Tripper sold for $385 in the 1973 catalog, and to this day it remains a standard in the industry, ideal for whitewater, leisure paddling, or canoe camping.

Despite success of OLD TOWN's fiberglass canoes and the new Royalex lines in canoes and kayaks, its big boat sales were slumping. While Gilman was busy

overseeing the production of the Royalex models, Deane Gray had remained focused on the big boats. The men involved in designing the powerboats— Cunningham, LaVoie, and others—tried to combat weak sales by offering more. From 1971 through '74, there were seven power boats developed: the magnificent 24-foot Atlantis sports cruiser; its little sister, the 20-foot Gull; the Eagle, also 20 feet but with fewer amenities; the Fisher, with a lower profile; and the Shrike, in both inboard and outboard options. Prices ranged from $11,175 for the top-of-the-line Atlantis to $2,195 for the affordable Shrike outboard. These intense efforts by the research and development team were still met by yearly market pressure for changes

and lackluster sales at the dealers, which put a constant strain on the OLD TOWN operating budget.

There was an additional problem with the big boats. It was always difficult to get big boats like the Atlantis out of Maine. Without the current interstate system, secondary roads had to be used and transportation laws prohibited trailers over 50 feet, which barely allowed carrying two of the powerboats at a time. The height of the trailer load was often taller than some underpasses would allow, and the route had to be carefully planned. On one particular trip, the driver deflated the trailer's tires and inched forward through an underpass hoping not to scrape his precious cargo. In other cases, special permits were needed for the load to proceed, delaying delivery. These re-occurring problems were highly frustrating for OLD TOWN and its dealers alike. With the added pressure of oil embargoes and the gasoline hikes throughout the '70s, Deane Gray made the difficult decision to drop the large boats from the line. By 1975 they no longer appeared in OLD TOWN's catalog.

Fully loaded, a delivery truck carrying an Atlantis and a Shrike makes its way down Middle Street. The staging that helped move the larger boats can be seen on the outside of the factory.
Old Town Canoe Company

[1]   Bart Hauthaway Interview.
[2]   K-1 is an abbreviation used to denote one man kayak; K-1W, one woman kayak; C-1, one man closed canoe; C-2, two man closed canoe; and C-2M, a "mixed" two-person closed canoe with a male and a female paddler.
[3]   Bart Hauthaway Interview, April 1997.
[4]   Lew Gilman Interview.
[5]   *Ibid.*

As fiberglass and Royalex canoe and kayak sales continued to grow at OLD TOWN, the company was quickly regaining the market prestige and respect it had lost due to competition from aluminum in the '50s and '60s. OLD TOWN was finally getting back on track, and everyone associated with OLD TOWN was optimistic about the future. While all these changes were developing, Deane Gray received a call from Sam Johnson of S. C. Johnson Wax fame. Although the boating industry knew that the OLD TOWN CANOE COMPANY had been struggling, it had never been offered for sale. But Johnson was looking into new ventures. An outdoorsman himself, Sam Johnson was diversifying his enterprises to include a line of businesses that catered to outdoor enthusiasts. He had already acquired Johnson reels, Minn Kota motors, Eureka tents, and other businesses that were associated with the outdoors, and he hoped to add OLD TOWN to the mix. The Gray family history actually paralleled his own in many ways. Back in 1886 in Racine, Wisconsin, his great-grandfather, Samuel Curtis Johnson, had started his family's business much like George Gray had done—through ingenuity, hard work, and perseverance. The Johnson heirs continued the tradition.

Sam Johnson was aware of the good times and bad that OLD TOWN CANOE had experienced and admired the fortitude of the Gray family. He also believes that "...every generation of a family business has to bring something new to the enterprise, something that hadn't been thought of by—and beyond the visions of—the previous generation."[1] He asked Deane Gray to fly to Racine for talks and although Johnson was a

15

The Big Comeback

Sam C. Johnson, the fourth genera-
tion of the Johnson family, whose
business began in 1886.
Old Town Canoe Company

development, the new ownership could continue all the innovations that had just begun.

Convinced that it was the best move for everyone, the Gray family sold the OLD TOWN CANOE COMPANY to Johnson Diversified in December 1974. A few days before Christmas, a letter posted on the company bulletin board notified the workers of the buyout. Rumor ran rampant as the workers worried about their futures. Shortly after the holidays, Sam Johnson and several of his top executives came to visit the plant and spoke with the workers. The group was apprehensive at first, but found Johnson to be highly convincing as he reassured them that few things would change. Deane told the workers about Johnson's own history in business with a family-owned operation similar to the Grays'. A tour of the facility followed. As the staff and crew individually met Johnson, each became more confident that this was the best move for the company.

As Johnson prepared to leave, he reassured everyone that things would be run as they had been, and he promised that as long as he owned the business, they "would continue to make wood-canvas canoes," the foundation of the OLD TOWN CANOE COMPANY.[2] Work on wood-canvas canoes still continues in the basement of the factory today, where approximately sixty wood canoes are crafted annually.

Following the Johnson buyout, things continued pretty much as usual. New record-keeping systems were initiated and a computer was installed to link sales records to computer stations in Racine, Wisconsin, worldwide headquarters for Johnson Wax. With Johnson's capital it was possible to offer dealers

fine host, Gray made no commitment. Gray did not discuss Johnson's proposal with any of his management or his workers. Finally, at the family's Thanksgiving gathering of 1974, Deane discussed Johnson's offer with other family members and recommended that they sell the business. After much discussion and with a clear understanding of the ups and downs of the canoe business, the family supported Deane's decision to sell. The family business would be passed to another family with similar ideals and a history of caring for its workers, customers, and community, and Deane could remain in his current position for as long as he chose. With Johnson's capital to back new

better incentives in the spring and fall buying programs. With the preseason orders initiated through the buying programs, material needs were projected more accurately and work schedules could be made more consistent.

Lew Gilman enjoyed meeting Sam Johnson and found him totally approachable and very likable. Above all, he felt that Johnson had a great deal of common sense, a quality Lew valued. Johnson, in turn, was impressed with Gilman's ingenuity and applauded his creativity. He hoped Lew would stay on to continue any endeavors he might choose to explore.

Royalex canoes were selling well enough to add three new models to the line, but Lew had been frustrated with Uniroyal's monopoly. Problems with quality and supply had angered Deane Gray enough for him to no longer use the Royalex name.[3] By 1975 Royalex was only referred to by its chemical abbreviation ABS in the OLD TOWN catalog. Gray was adamant that the Uniroyal company would get no further advertising from OLD TOWN CANOE. In 1976 Gray began calling the material "Oltonar" for OLD TOWN Royalex.

Gilman had been working on a new idea that did not involve Royalex. He was tired of being dependent on one supplier and was attempting to fashion a material similar to Royalex out of polyethylene. He did extensive research, sought out suppliers, and then began to experiment with rotational molding and the different densities of polyethylene. Ever practical, he geared up by hauling a kitchen range into the factory, then walked to the corner hardware store and bought some bolts, fastenings, and two pie plates. After fash-

The faces of the employees express their concern as they listen to the details of OLD TOWN CANOE COMPANY's acquisition by Johnson Wax.
Old Town Canoe Company

Royalex, now called Oltonar by OLD TOWN CANOE, was fashioned into many different models. This Kennebec model was built with rocker designed especially for whitewater use.
Old Town Canoe Company

ioning a mold out of the two pie plates, he constructed rotational molding equipment from some old gears, leftovers from the days when OLD TOWN CANOE was a Johnson motor dealer. The rotation was crucial, for the mold had to rotate on two axes at the same time. Lew poured some of the polyethylene powder into his pie plate mold, placed it in the oven and watched it rotate. He then added the second powder for the second layer, and then finally the third. The time, the amount, and the compounds were all tested by trial and error, like a chef concocting a new recipe. At times it was highly frustrating, as the molten polyethylene did not always spread evenly in the plates. At other times it cooperated and flowed smoothly, producing the chemical bond desired and boosting Lew's hope that he was on to something.

After the Johnson buyout, Lew was allowed to continue his experimentation. He even built a full-scale rotational mold which produced canoes, but its performance was irregular at best. Because a two-part mold was used, early canoes often had bubbles at the keel line where air had fizzled in or out of the mold. It was a difficult problem to correct. Lew's frustration grew when Deane Gray, his foremost supporter, left the company in 1978 and was replaced by a Johnson general manager who was much more skeptical of rotomolding. Despite Lew's protests, work on rotational molding was halted. No more time and money was to be put into the idea. Disillusioned and disappointed, Gilman walked away from the project.

At about the same time, the company tried to reach out in other directions. With the loss of Bart Hauthaway as OLD TOWN's chief kayak designer in the early '70s, kayak design and innovation had stagnated—and sales reflected it. If OLD TOWN was to remain in the kayak business, something new had to happen. In an attempt to re-enter the market, in March 1979 OLD TOWN CANOE COMPANY acquired sole United States distribution for Lettmann and Prijon kayaks.[4] Klaus Lettmann and Toni Prijon were former world champion racers, and their designs had been paddled to world championships. OLD TOWN hoped its investment would pay off in the following year when the World Whitewater Championships would be held for the first time in North America, in the province of Quebec. The defending world championship team from West Germany would be using kayaks designed by Lettmann and Prijon; now the U.S. team would also have that choice. OLD TOWN hoped that the World Championships would spur a kayak-buying spree in America and in Europe. This was only one of several moves to further popularize the company.

Another innovation was to introduce a canoe that could be easily stacked, saving valuable factory space and time. The Carleton, as the model was called, was available to the customer in kit form. (This new model should not be confused with the earlier Carleton fiberglass line of canoes or the wood-canvas Carletons.) Constructed of Royalite, a product developed by Uniroyal researchers, it had an ABS solid core instead of foam. Royalite had "memory" that would allow a twisted hull to pop back into shape with a few kicks in the right place. Gunwales were specially made for this canoe, and it featured a keelson of extruded aluminum. Flotation was added to each end of the canoe, and the canoe was given a baked urethane finish. The

In 1976, when I first saw the old green canoe leaning against a house by Eagle Springs Lake, Wisconsin, I was immediately attracted to it. Even at a distance, its classic lines were appealing—different from any canoe in my experience. I had paddled lots of synthetic canoes and owned a few, but never a "woodie." I wondered if anyone used it or cared for it anymore.

I was later introduced to its owner, who had been saving the old family canoe, hoping his grown son would take possession and responsibility for it. I told him of my interest in canoeing, my wish that classic wooden canoes should be cared for, and my willingness to care for his, if it needed a new home. After some thought, he gave it to me.

The canoe's poor condition made the deal not the windfall it might seem to be. Its time resting on the ground had had the predictable effect of causing rot in the gunwales, rib ends, and decks. I took it home to Chicago, planning its restoration and thrilled to be the owner of an Old Town regardless of its condition. I sent the serial number to the Old Town Canoe Company, and they responded that I had a 1937 OTCA model.

I gathered materials for the restoration, hindered by my meager grad-student's income, and began the project in earnest in the summer of 1978, after I graduated. I worked in a friend's back yard. When fall came, we hung the sad-looking stripped hull from the rafters of his garage until the return of spring made working outside feasible again. I told myself it was the safest, best storage the canoe had ever had.

But a few months later, the infamous blizzard of 1979 dumped 24 inches of wet snow on Chicago. The garage collapsed, along with over 100 others around Chicago, and the canoe was reduced to kindling. My friend's suffering blunted my self-pity. He had kept his two Alfa Romeo sports cars in the same garage. My loss seemed too small to talk about, but privately I grieved. I still do.

—*Kim Apel, San Clemente, CA*

Carleton was OLD TOWN's answer to Coleman, whose models made of Ram-X were stealing a large percentage of the lower-priced market. Twenty thousand Coleman canoes were being sold annually, compared to OLD TOWN's 5,000. Although the Carleton proved competitive in the marketplace, OLD TOWN's management was disappointed with Royalite's inconsistent performance. The hulls often warped, and customers seemed reluctant to buy an OLD TOWN with an interior aluminum keel. Royalite was a recycled product, which was certainly commendable, but the technology still needed improvement. Although there are still healthy Carletons in service today, on the whole the Carleton model did not live up to OLD TOWN's quality standards.

Lew Gilman saw the Carleton demise coming and hoped his rotational molding idea might be given another chance. In the interim, new materials had been developed that were more consistent and predictable. OLD TOWN's new general manager, John Blass, gave the okay and Lew was again allowed to continue his experimentation. By 1983 Gilman's experimentation paid off and he finally perfected the process to produce a three-layer canoe hull consisting of an outer layer, a foam inner core, and a resilient inner layer, all of polyethylene. He went on to mold the new generation of OLD TOWN canoes. The canoes, christened Discovery after the space shuttle, were first sent to livery and rental outfits across the country for a full year of testing. Every year of rental use is equal to ten years of normal wear and tear, and this was a good opportunity to see if the material could withstand the demands of its customers.

Outfitters were enthusiastic. The hulls held up well and by 1985 the 17-foot 4-inch Discovery made its debut in the catalog.[5] Polyethylene was not a new material, but Gilman's CrossLink3 rotational molding

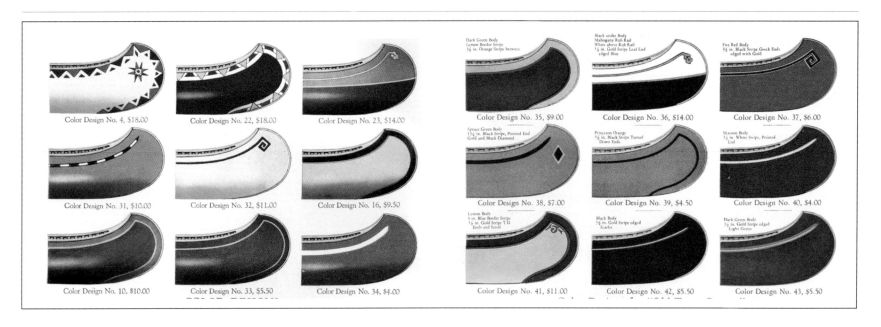

Color Design No. 4, $18.00   Color Design No. 22, $18.00   Color Design No. 23, $14.00

Dark Green Body / Lemon Border Stripe / ⅜ in. Orange Stripe between

Black under Body / Mahogany Rub Rail / White above Rub Rail / ⅛ in. Gold Stripe Leaf End edged Blue

Fire Red Body / ⅜ in. Black Stripe Greek Ends edged with Gold

Color Design No. 35, $9.00   Color Design No. 36, $14.00   Color Design No. 37, $6.00

Color Design No. 31, $10.00   Color Design No. 32, $11.00   Color Design No. 16, $9.50

Spruce Green Body / 1⅝ in. Black Stripe, Pointed End / Gold and Black Diamond

Princeton Orange / ¼ in. Black Stripe Turned Down Ends

Maroon Body / ½ in. White Stripe, Pointed End

Color Design No. 38, $7.00   Color Design No. 39, $4.50   Color Design No. 40, $4.00

Color Design No. 10, $10.00   Color Design No. 33, $5.50   Color Design No. 34, $4.00

Lemon Body / 3 in. Blue Border Stripe / ½ in. Gold Stripe T.D. Ends and Scroll

Black Body / ½ in. Gold Stripe edged Scarlet

Dark Green Body / ½ in. Gold Stripe edged Light Green

Color Design No. 41, $11.00   Color Design No. 42, $5.50   Color Design No. 43, $5.50

Right: Lew Gilman's development of Cross-Link3 rotomolding led to a complete line of canoes christened with the name "Discovery."
Old Town Canoe Company

process was, and it later received two patents in the United States and one in Canada.[6]

The quiet introduction of the Discovery in 1984 had been a turning point for OLD TOWN. The public responded enthusiastically to a canoe with great styling, a versatile, strong, resilient hull, and an affordable price. For the first time since the company was sold in the mid-'70s, the company found itself in the black.[7] By the end of 1985, OLD TOWN had already sold 5,000 units and production could not keep up with demand. The company's sales grew 27 percent that year, and it regained its title as the largest manufacturer of canoes in the world. Coleman, who had dominated the market for several years, suffered a sales slump of 20 to 30 percent during the same two-year period.

In 1985 a second canoe was added to the Discovery line, a 15-foot 8-inch model called the Discovery 16.

Above: An unbelievable amount of information is contained within these small storage boxes. Collectively, they tell the history of the OLD TOWN CANOE COMPANY.
Old Town Canoe Company

Facing page, above: Treasures in the records room include information about the various design motifs that were available. The old catalog pages from the early 1930s show over fifty choices. Old Town Canoe Company

Boxes of History

Tucked in a space at OLD TOWN hardly the size of a closet, where a bare light bulb dimly illuminates the shelves, are hundreds of wooden boxes measuring approximately 6 by 8 by 14 inches, each filled with hundreds of cards that contain a wealth of information. Collectively the cards describe the OLD TOWN CANOE COMPANY's manufacturing history. Each card details vital information about one canoe or boat constructed, including its serial number, boat model, grade, when it was planked, canvassed, railed, filled, and painted. Many a handwritten note names the craftsman or gives a description of the customer's requests for choice of materials, special paint designs, or unique embellishments.

There are four series of boxes within the tiny room. The OLD TOWN canoe series begins with card #1604 dated April 25, 1906. The card describes a dark green, 18-foot Charles River model in CS grade, with the name *Mayada* painted on the bow. It sold to G. E. Graham of Indianapolis, Indiana.

Although on the surface the cards look relatively complete, there are large gaps in the series and some inconsistencies in numbering. The earliest years of the company are missing. A few cards with numbers prior to #1604, and some after that number, are contained in a miscellaneous box which holds records of canoes brought back for repair.

In addition to the sequence of OLD TOWN canoes, there are separate runs for Carleton canoes, OLD TOWN flat-bottomed boats, and OLD TOWN rowing skiffs. The Carleton cards show an assortment of numbers but become more consistent with the 7700s following the acquisition of the Carleton company by OLD TOWN in March 1910. Flat-bottomed utilitarian boats produced until 1957 have a separate numbering system beginning with #1 in March 1927. Rowing skiffs are also tracked starting in 1941.

The oaktag cards are dog-eared from handling, as employees keep up with many requests for research information based on the serial numbers. For a three dollar charge (as of this writing), a serial number will be researched. The card will be photocopied, and along with additional pertinent model information, will be given to the inquirer. New technology allows the cards to be scanned into a computer database for future research. Eventually those parties interested will be able to analyze the captured data to see the ebb and flow of a century of canoe production.

To alleviate confusion, the company adopted a new identification system. Canoes would now be designated by name and length. A new 16-foot 9-inch model was soon introduced as Discovery 169. This canoe, sized between the newly named Discovery 158 and the Discovery 174, was fashioned after Gilman's legendary Royalex Tripper, with a deep-V entry and a high-volume, durable hull ready to take on whitewater, wilderness tripping, or family paddling and picnicking. Within five years of its introduction, the Discovery line gave OLD TOWN 25 percent of the market, compared to the 10 percent it had held previously.[8]

Gilman's Discovery was probably the most innovative success to ever come out of a pie plate! Within a short decade, the Discovery line grew to include a solo canoe that weighed a mere 43 pounds and was ideal for the sports lover, another marketed to rowing enthusiasts, and two that retained the classic lines of the Tripper and the Penobscot. Gilman's Discovery canoes have a strong, solid track record and have proven themselves with years of customer satisfaction. As with many discoveries, it came from a man who had been told, "It can't be done."

As a way of celebrating its tenth anniversary, the Discovery was put to the same test that its predecessor, the Royalex canoe, had endured years earlier — a drop from the factory roof. As Scott Phillips, a member of the research and development team recalls, he and a co-worker made their way to the edge of the factory roof with some trepidation, not because they doubted the Discovery's durability but because the height made both men nervous. They stepped up to the edge and gave the boat a heave. It plummeted downward toward the broken asphalt below. Like its predecessor the Royalex Tripper, the Discovery 169 bounced, folded slightly, and lunged forward with the bow taking the brunt of the fall. The nose was scratched reasonably for the contact made, but there were no signs of damage anywhere on the hull. It survived successfully and its two greatest admirers were the men who had released it.

Another milestone occurred on October 31, 1984, when OLD TOWN CANOE acquired the White Canoe Company, merging two companies with long heritages. At the time, White Canoe was the oldest continuously operating builder in the country (1888), and its path had crossed OLD TOWN's often. It first employed OLD TOWN's premier builder, Alfred Wickett, and later provided OLD TOWN CANOE with Walter King, a former owner of the White company, who led OLD TOWN CANOE's fiberglass revolution.

By merging the two businesses, OLD TOWN hoped to double its fiberglass and Kevlar production; the move also enhanced the marketing prospects of White's Stillwater lines via OLD TOWN's national and international distribution channels.

In 1989 at the centennial celebration of White Canoe Company, OLD TOWN issued a commemorative canoe, an 18-foot 6-inch cedar-planked Centennial model, patterned after White's famous 18½-foot "Guide." The canoes were built by Jerry Stelmok of Island Falls Canoe Company of Dover-Foxcroft, Maine, considered to be the keeper of the White legacy. The Centennial canoes captured the features that distinguished White's products: his trademark

D-shaped stern seat of hand-caned bentwood, cedar ribs and half ribs in the floor, gunwales that extended slightly beyond the stems, brass wrap-around stem plates to protect the canvas seams from abrasion, and a protective shellac coating on the bottom. Finishing touches included mahogany decks and outwales, hand-carved white ash thwarts and carrying yoke, and a brass commemorative plaque indicating the serial number of this limited-edition canoe, a canoe to be admired, used, and cherished. Only 25 were produced for the anniversary, in a color choice of rich dark blue or deep forest green paint. Canoe number one is on display in the OLD TOWN CANOE factory store in Old Town.

One year following the celebration, the White line was incorporated into OLD TOWN CANOE's offerings. Management felt the two companies competed for the same market, and the cost of maintaining separate companies presenting similar products was no longer feasible. The fiberglass lines from White were a welcome addition to OLD TOWN's fiberglass offerings.

The success of any great company depends upon its ability to respond to changes in the buying public. Recognizing a rising interest in kayaks, OLD TOWN re-entered the market in 1990. Through a partnership agreement with Plastiques LPA, Ltd., of Mansonville, Quebec, OLD TOWN would act as United States distributor of Plastiques' line of sit-on-top kayaks. Its first Dimension kayak featured an open cockpit with contoured back and leg support for all-day paddling comfort. It was constructed of tough, rotomolded polyethylene, and offered paddlers a convenient dry storage compartment as one of its features. Other

models followed. The Dimension line was offered through a separate catalog and did not carry the OLD TOWN name. John Blass, general manager of the OLD TOWN CANOE COMPANY, felt these kayaks were a good concept and was pleased with their appeal, stability, and quality. Neither OLD TOWN nor Plastiques expected the outpouring of interest in kayaks that ensued. The popularity of kayaking was rising yearly, and there was a demand for kayaks carrying the OLD TOWN name.

Like the Grays had done so many times before, OLD TOWN responded to the market. In 1995 it unveiled the Otter, a 9-foot 6-inch kayak constructed with a single layer of polyethylene, designed and produced by OLD TOWN.

It had been ten years since a kayak had displayed the OLD TOWN name. During that hiatus, the kayak industry had broadened its market beyond whitewater and Olympic kayakers. It now appealed to a wider spectrum of recreational paddlers with new designs that were stable and easy to manage. For the canoeist who was tired of trying to find a partner, frustrated with putting a canoe on the top of a vehicle, or was irritated by the effects of the wind on the water, the kayak seemed to be the answer. As interest increased, sales followed. OLD TOWN responded with more specialized models. In 1998 OLD TOWN kayaks are projected to account for 42 percent of the company's unit sales.

In an effort to make the kayak more durable, with greater flotation, the research and development team at OLD TOWN developed PolyLink3, a linear polyethylene with a tri-layer construction, the middle layer

Right: A variety of sit-upon kayaks, marketed under the Dimension name, are available from Plastiques LPA, Ltd., now a subsidiary of OLD TOWN CANOE.

Far right: Leisure Life, Ltd., now another part of the OLD TOWN family, offers products that are fun for old and young alike. Old Town Canoe Company

Below left: Like his father Walter, before him, Geoff King is now head of Research and Development at OLD TOWN CANOE. Here he's fishing from one of his most recent designs, the Loon I, now called the Loon 138.

Below right: The largest maker of sit-upon kayaks is Ocean Kayak from Ferndale, California, one of the latest additions to the OLD TOWN family. Old Town Canoe Company

viding the buoyancy.[9] This material eliminated the need for adding the foam blocks or air bags necessary in a single-layer polyethylene model. It also made a stiffer hull that was lighter in weight. This technology is now being used on the Loon and Millennium series, the Egret and Heron kayaks, and on the new Guide Canoe.

"Like father, like son," can certainly be applied to Geoff King, now head of research and design at OLD TOWN. He got his start working at White Canoe as a teenager. At college he furthered his education, learning the intricacies of design and the techniques involved in working with high-tech materials. He followed his father Walter's footsteps, coming to OLD TOWN in 1969 and performing a variety of tasks since then. Recently Geoff and his crew have introduced new models at OLD TOWN, from a rowing skiff to sport boats, from canoes to the current Millennium series of kayaks. It's Geoff's work on the molds that produces the beautifully finished products seen on the water. Geoff is proud to be connected with OLD TOWN, and living proof that family legacies continue there.

In the summer of 1997, under its parent company Johnson Worldwide Associates (JWA), OLD TOWN substantially increased its line of kayaks by acquiring Ocean Kayak of Ferndale, Washington. Ocean Kayak—the largest kayaking manufacturer of sit-on-tops—was founded by Tim Niemier. Niemier first developed his kayaks as a way of building a better mousetrap—a kayak that anyone could use, one that was affordable, portable, durable, and safe.

Like so many before him, he began building in his garage. He started with fiberglass, but by 1985 he had

Left: The old masters, like Emedy Baillargeon, will always have that touch. Old Town Canoe Company

Below: Wood: past, present, and future. Old Town Canoe Company

switched to the more durable and cost-effective polyethylene. His first model was a simple unit called a Scupper, a self-bailing kayak that had holes for drainage incorporated into the design. From there he went on to develop models for more specific purposes, including models designed for fishermen and skindivers. Although part of the JWA family, Ocean Kayak operates independently and broadens the scope and appeal of the watercraft lines.

With JWA's backing, Ocean Kayak enjoys new distribution opportunities through the 800 OLD TOWN dealers worldwide. Specialty stores, including the export market, make up 79 percent of OLD TOWN's distributorship. The marriage of Ocean Kayak and OLD TOWN has created the largest kayaking manufacturing effort in the world, and it's still growing.

In September 1997, Plastiques LPA, Ltd., which had been affiliated with OLD TOWN since 1990, was purchased from co-founder Pierre Arcouette. Arcouette and his company have been instrumental in helping to advance the rotomolding process. This acquisition brought Arcouette's expertise to the company and increased OLD TOWN's manufacturing capacity, introducing state-of-the-art kayak manufacturing equipment.

Realizing that the watercraft market has a broad scope, another acquisition took place in February 1998 with the newest addition to the watercraft division:

"...I had a vision of a company that would make high quality yet affordable marine products that appeal to consumers wherever they used water for recreation...," said Charlie Billmayer, founder and CEO of Leisure Life, Ltd., as he talked about his early beginnings making pedal boats back in Michigan in 1977. His sentiments echo those of John Blass, Sam Johnson, the Gray family, and all those past and present who are connected with the OLD TOWN CANOE COMPANY. Leisure Life, Ltd., which is one of the largest small-boat manufacturers in the world, produces pedal boats, fishing boats, dinghies, canoes, and the environmentally friendly new ElDeBo Electric Deck Boat, ideal for places where gasoline motors have been barred. With foresight and careful acquisition policies, JWA has seen its watercraft division quadruple its sales from 1993 to 1998.

Even with the changes in ownership and the introduction of the high-tech plastics and new technology, the climate at the OLD TOWN CANOE COMPANY hasn't changed a lot. The tradition continues with every canoe, boat, and kayak it produces. You can sense the ghosts as you walk throughout the plant. They are there: the craftsmen, the loggers, the teamsters—they will always be present. OLD TOWN has withstood the test of time—one hundred years makes it the most famous and revered "granddaddy" of them all. OLD TOWN has left an enduring legacy. There is magic when an early OLD TOWN canoe is discovered under a camp porch or in an old barn, waiting to be restored and rechristened in the water again. In the minds of countless paddlers—wilderness trippers, ocean kayakers, kids at summer camps, families on outings, whitewater enthusiasts, sportsmen, race competitors, photographers, naturalists, those who enjoy a romantic shoreline picnic, or a grandparent taking his "legacy" fishing along a meandering stream—OLD TOWN offers the magic carpet.

...Now when it's time for me to die
And take my last long ride
Old Town Canoe if you stand by
I'll reach the other side.
—Anonymous

[1] Samuel C. Johnson, *The Essence of a Family Enterprise* (Indianapolis, Indiana: The Curtis Publishing Company, Inc., 1988).

[2] Sam Johnson Interview, July 1996.

[3] Lew Gilman Interview.

[4] David Cheever, "Old Town Canoe Has New Kayaks," *Bangor Daily News*, March 21, 1979.

[5] OLD TOWN CANOE COMPANY later named this canoe the Discovery 17, and then renamed it the Discovery 174.

[6] CrossLink3 is a patented process designed by Lew Gilman.

[7] John Blass, General Manager, OLD TOWN CANOE COMPANY, Interview.

[8] *Ibid*.

[9] PolyLink3 is a patented process designed by Lew Gilman.

Note: "My Old Town Canoe" on the prologue page and this page is from the *Old Town Enterprise*, December 25, 1921.

FROM THE CATALOGS

BIBLIOGRAPHY

INDEX

I. F. Model

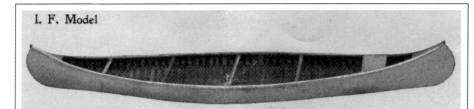

This is a special Indian model and was planned by one of our Indian workmen to meet the requirements of the professional guides and of sportsmen who need a canoe for use in the still water of pond and river and also for poling over rapids and through quick-running shallow water. Is of generous width, flat enough on the bottom to have good carrying qualities and draws but little water. At bow and stern it narrows sharply, is quick in getting under way.

F. B. Model

Canoes from this model are made in one length only — seventeen feet — but in any grade. It is rather flatter upon the bottom, is fairly sharp at stem and stern, and has good bilge. This is a very good shape for general work and will meet the wants of those preferring a flat bottomed canoe.

G. G. Model

This canoe is well adapted for all around work. It is similar in outline to our I. F. Model but is not quite so sharp at bow and stern and has more bilge. For sporting work this is a good shape. It is about half way between our H. W. and our I. F. Models. Made eighteen feet long only.

Guides' Special

This canoe is planned throughout for the use of professional guides and sportsmen.   It is strongly built of good lumber; has cedar ribs and planking — the planking is in long lengths, thus making a stiff canoe; — has Spruce Gunwales and Outwales, White Ash Decks, Thwarts and Posts.   Is copper and brass fastened.   Has canvas or caned stern seat.   No bow seat furnished unless ordered.   Is painted under canvas.   Has No. 6 canvas which is carefully filled with white lead and oil only, well rubbed down.   Inside is filled, rubbed down and varnished.   An honestly built canoe, put together for wear, guaranteed satisfactory.   No money is put into ornamenting.

Length	Width	Depth	Price
18 feet	33 1-2 inch.	12 1-2 inch.	$26.00
19 "	34 1-2 "	13 "	27.00

Memo. : We will build to order only, at same prices as above a very sharp canoe having otherwise the same lines as our I. F. Model.

Special Canoe for Salmon Fishing

We also build in this grade and on this model a canoe planned for salmon fishing.

Length 20 feet,   Width 36 inches,   Depth 13 inches,   Price, $30.00.

CANOE PRICE LIST

	Length	Width Amidship	Depth Amidship	Grade A. A.	Grade B. B.	Grade C. S.
Model H. W.	15 ft.,	29 in.,	11 in.,	$33.	$30.	$25.
	16 "	30 "	11 1-2 "	35.	32.	25.
	17 "	31 "	12 "	38.	34.	26.
	18 "	32 "	12 1-2 "	41.	37.	26.
Model I. F.	15 ft.,	30 1-2 in.,	in.,	$33.	$30.	$25.
	16 "	31 1-2 "	11 1-2 "	35.	32.	25.
	17 "	32 1-2 "	12 "	38.	34.	26.
	18 "	33 1-2 "	12 1-2 "	41.	37.	26.
	19 "	34 1-2 "	13 "	44.	40.	27.
Model F. B.	17 "	32 "	12 "	$38.	$35.	$26.
Model G. G.	18 "	35 "	12 "	$41.	$37.	$26.

Keel put on, if ordered, at no extra cost. Outside stems $2.00 extra. Long decks $5.00 to $10.00 extra. Mahogany outwales $2.00 extra. Oak or Birch outwales $1.50 extra.

Names will be put on for 12 cents a letter. Gold stripe for $2.00. Special decorations to order.

Terms

Cash with order or satisfactory reference. Upon special order work a payment on account will be required. On C. O. D. Express shipments part payment will be required in advance.

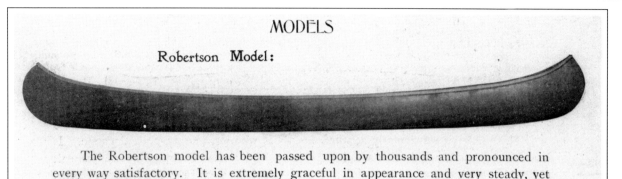
MODELS

Robertson Model:

The Robertson model has been passed upon by thousands and pronounced in every way satisfactory. It is extremely graceful in appearance and very steady, yet with the fine lines that give speed and ease of paddling. As a model for pleasure canoes this cannot be excelled.

Canvas Covered Dinghy or Yacht Tender

DESIGNED BY B. B. CROWNINSHIELD, NAVAL DESIGNER, OF BOSTON.

This is a very light and satisfactory tender for a launch or yacht. It will carry a large load, is easily handled and tows well. Has cedar ribs and planking; stems, knees, rubbing streaks, rudder, etc., are of Ash or Oak; is copper fastened; wood work finished natural color and varnished. The covering is of heavy canvas, thoroughly filled with a waterproof coating and painted. The construction is the same as in our canoes but heavier material is used.

Length 11 1-2 ft.   Price $46.00, including rudder, other attachments extra
     "   12 1-2 ft.   "   50.00,   "   "   "   "   "
     No. 1 Brass Swivel Rowlocks like cut $2.40 per pair
Extra sockets for above $.80 per pair   Bilge Keels put on for $1.00

Charles River Model (formerly called
Robertson Model)
OLD TOWN CANOE COMPANY
1903 Catalog

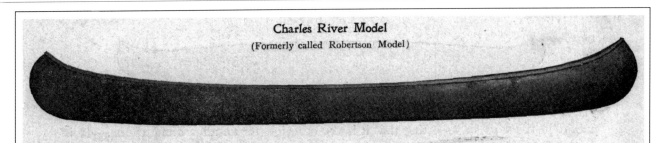

Charles River Model
(Formerly called Robertson Model)

This canoe is designed particularly for use on rivers and small ponds. It is extremely graceful in appearance and very steady. combining beauty in outline with speed and ease of paddling. The sides have a little more bilge or convexity than do those in our H. W. model, and it is also straighter on the bottom from bow to stern. It is an ideal model for a pleasure canoe.

Canvas Covered Double End Boat
OLD TOWN CANOE COMPANY
1906 Catalog

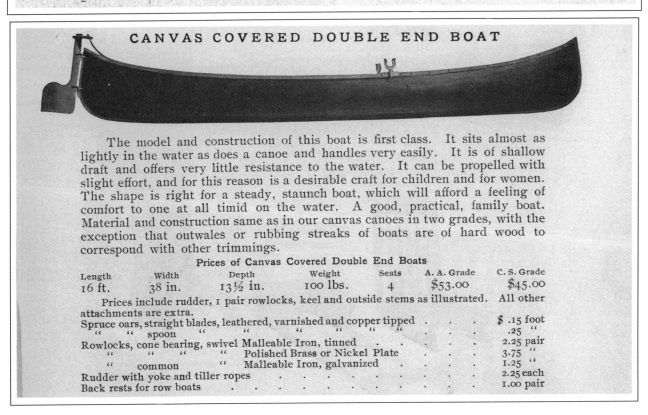

CANVAS COVERED DOUBLE END BOAT

The model and construction of this boat is first class. It sits almost as lightly in the water as does a canoe and handles very easily. It is of shallow draft and offers very little resistance to the water. It can be propelled with slight effort, and for this reason is a desirable craft for children and for women. The shape is right for a steady, staunch boat, which will afford a feeling of comfort to one at all timid on the water. A good, practical, family boat. Material and construction same as in our canvas canoes in two grades, with the exception that outwales or rubbing streaks of boats are of hard wood to correspond with other trimmings.

Prices of Canvas Covered Double End Boats

Length	Width	Depth	Weight	Seats	A. A. Grade	C. S. Grade
16 ft.	38 in.	13½ in.	100 lbs.	4	$53.00	$45.00

Prices include rudder, 1 pair rowlocks, keel and outside stems as illustrated. All other attachments are extra.

Spruce oars, straight blades, leathered, varnished and copper tipped . . .	$ .15 foot
" " spoon " " " " " " .	.25 "
Rowlocks, cone bearing, swivel Malleable Iron, tinned	2.25 pair
" " " " Polished Brass or Nickel Plate . . .	3.75 "
" common " Malleable Iron, galvanized . . .	1.25 "
Rudder with yoke and tiller ropes	2.25 each
Back rests for row boats	1.00 pair

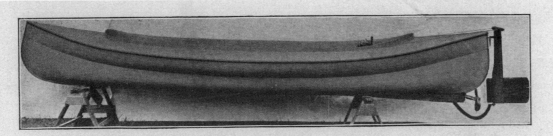

POWER CANOE BOATS

Power canoes have been thoroughly tried out in this vicinity and proved to be in every way satisfactory. As we build them they have some points of decided superiority, such as light weight, light draft, safety and beauty.

CONSTRUCTION

The hull is built of selected Maine cedar, the ribs being ½ inch thick, 3 inches wide, placed 1 inch apart; the planking is $\frac{5}{32}$ inches thick, fastenings are all copper, brass or galvanized iron. The hull is covered with heavy No. 4 cotton duck put on in one piece, which is thoroughly filled and brought to a smooth, glossy surface, with a waterproof composition; the inside of the hull is strengthened and stiffened by braces extending from stem to stern.

SHAPE

The design of the hull is identical with that of our XX canoe, made deeper and having straight sides. This model has a flat floor, giving stability and sharp lines forward and aft and is planned for speed.

DIMENSIONS

Length   18 feet   Depth   19 inches
Width inside, amidship   37   "
    "   over all   "   49   "

WEIGHT

Hull complete, 300 pounds; engine and all fittings complete, 250 pounds; total, 550 pounds; about one half that of the ordinary heavy all wood boats.

CAPACITY

Four passengers comfortably, five may be carried.

Outside the boat is painted dark green or any color desired, and varnished with spar varnish. Inside the finish is natural wood in varnish.

AIR CHAMBERS

On each side are built sponsons or air chambers, 10 inches deep, 5½ inches wide amidship tapering forward and aft, and having a rounded top for carrying off water. These air chambers add very much to the desirability of this boat, and are a decided improvement over the ordinary air tanks sometimes placed at bow and stern under the decks, which while making a boat non-sinkable do not add to its stability. These side sponsons in addition to increasing very materially the buoyancy of our canoe make it difficult to capsize and correspondingly safer for a small power boat. A 2-inch coaming runs completely around the inside of the boat.

SEATS

The boat may be equipped with our folding caned chairs; four going in very comfortably. Chairs are movable and may be changed about to suit the passengers and to trim boat right.

RUDDER

Rudder and yoke are of oak with a braided tiller rope which is passed through brass eye bolts around inside of boat.

KEEL

An oak keel forward, 1 inch deep, increasing from center to 3 inches deep at stern to carry propeller shaft and stuffing box, with a heavy iron skeg to protect propeller blades.

POWER

We will supply this boat fitted with either a ¾ H. P. or a 1½ H. P. gasoline engine. With the smaller engine a speed of about six miles an hour may be attained, while the larger engine will give about eight to nine miles an hour. Generally we furnish an engine carrying the supply of gasoline in the cast iron base. This does away with piping for the gasoline and the possibility of leaky valves and joints. When desired we will supply an engine without the tank base, with gasoline tank under the forward deck piped to engine. This engine is simple in construction and thoroughly built from templates, thus all working parts are interchangeable. Cylinder and head are water jacketed and are cast integral. Connecting rods and bearings are of phosphor-bronze. Crank shaft is of steel, drop forged. Piston is accurately turned and fitted with two cast-iron rings.

ENGINE

We have selected for our boats a well-built engine—not the cheapest on the market. We can buy one for about one-half what we pay now, but do not believe the cheap engine will give our customers satisfaction. Will instal any engine a customer may prefer.

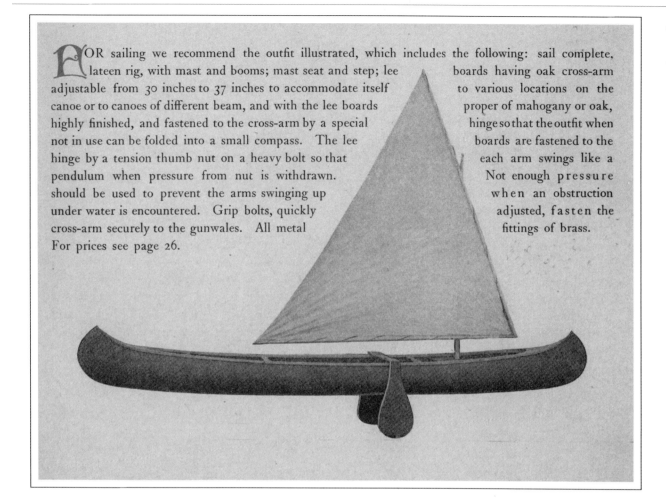

ᖴOR sailing we recommend the outfit illustrated, which includes the following: sail complete, lateen rig, with mast and booms; mast seat and step; lee adjustable from 30 inches to 37 inches to accommodate itself canoe or to canoes of different beam, and with the lee boards highly finished, and fastened to the cross-arm by a special not in use can be folded into a small compass. The lee hinge by a tension thumb nut on a heavy bolt so that pendulum when pressure from nut is withdrawn. should be used to prevent the arms swinging up under water is encountered. Grip bolts, quickly cross-arm securely to the gunwales. All metal For prices see page 26.

boards having oak cross-arm to various locations on the proper of mahogany or oak, hinge so that the outfit when boards are fastened to the each arm swings like a Not enough pressure when an obstruction adjusted, fasten the fittings of brass.

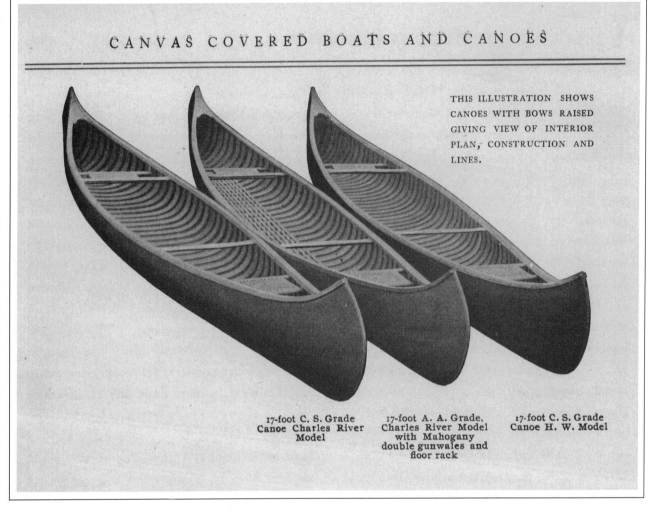

CANVAS COVERED BOATS AND CANOES

THIS ILLUSTRATION SHOWS
CANOES WITH BOWS RAISED
GIVING VIEW OF INTERIOR
PLAN, CONSTRUCTION AND
LINES.

17-foot C. S. Grade
Canoe Charles River
Model

17-foot A. A. Grade,
Charles River Model
with Mahogany
double gunwales and
floor rack

17-foot C. S. Grade
Canoe H. W. Model

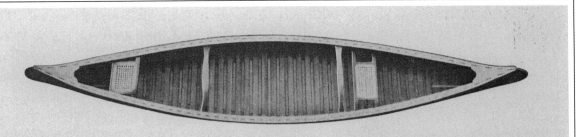

" OLD TOWN ' IDEAL ' CANOE " (OPEN GUNWALES AND HALF RIBS.)

16 ft. A. A. C. R.,	$38.00	17 ft. A. A. C. R.,	$40.00
Open mahogany gunwales,	3.00	Open mahogany gunwales,	3.00
Half ribs,	2.00	Half ribs,	2.00
Keel,	1.00	Keel,	1.00
	$44.00		$46.00

IF you wish a canoe with distinctive features this is built particularly to fill your want. It is the canoe without a peer, designed for the discriminating purchaser, and assures him exclusive ideas in canoe construction.

Open gunwales of mahogany and half or short ribs give originality, beauty and utility. The open gunwales accomplish ease in washing out where, in canoes of regular construction, dirt gathers beneath the gunwales when the canoe is turned over to remove water, etc., and is cleaned out with difficulty if at all. Half ribs add strength to the bottom of the canoe, making a floor rack unnecessary. Their lengths are graduated from amidships to the ends of the canoe and conform to the flatness of the bottom. The ends of the ribs as shown in the illustration are cut off square and held in place by two gunwales which are screwed and dowelled together, the rib ends being finished flush and smooth with the top surface. A small pocket is made in the lower inner edge of the outside gunwale to receive the planking and canvas. This canoe will satisfy the critical purchaser and is cataloged by us this year for the first time. We shall carry it in stock only in A. A. grade, Charles River Model, 16 and 17-foot lengths, colored dark green. This construction can be accommodated to any canoe we make for the additional charges as noted, but three weeks time will have to be allowed for the work. *Get your order in early.*

Specify if keel is wanted.

Code Word, prefix Ge to word for canoe as given in price list pp. 8. Example : *Geramp* would be 16 A. A. C. R. Dark Green with Open Gunwales and Half Ribs.

"Old Town 'War' Canoe" (34=ft.)

DON'T you think these boys are getting all the pleasure possible from their summer outing? This picture represents one of our "War Canoes" on a lake in New Hampshire, and when fully manned it has passed every launch it has encountered for a short sprint. It is built from a special model, the planking and ribs are of extra thickness, it is reinforced longitudinally by floor braces, and the bottom is still further strengthened by a keel outside and a floor inside. All materials are carefully selected to procure the maximum strength. It is equipped with one stern seat for the coxswain and the thwarts (spaced 27 inches apart) are 4 inches wide, ample width for the paddlers to sit or rest on. Decks 30 inches long are installed and the sides are strengthened by heavy spruce "open" gunwales.

These canoes we build to order only, and four weeks' time is required for completion. The price includes crating and loading for shipment, and, as they are generally forwarded on open flat cars, we use great care in packing, so as to avoid any possibility of damage in transit.

We shall be pleased to correspond with you in regard to these canoes, and if the lengths do not fill your requirements, will submit quotations on any length canoe you may desire. Our experience, however, has shown these lengths to meet nearly every requirement and we have supplied them particularly for the use of canoe clubs.

Length	Width	Depth	Price
34 ft.	40 in.	14 in.	$135.00
25 ft.	44 in.	14 in.	75.00

"Old Town 'Otca' Canoe"

IN our "Otca" Model are comprised dimensions which are distinctive in our other models and the style of finish includes open spruce gunwales and 20-inch long decks with low combing. The bows are full as in our H. W. Model, the floor is flat like our Charles River Model, while increase of beam provides a maximum of steadiness without proportional loss of speed. It's a fast canoe. We know the model will make as many friends as it has users, while the originality in finish provides a canoe in C. S. Grade of relative attractiveness to our "Ideal Canoe." It was introduced last year and proved so popular that we had difficulty in filling orders. This season our stock will be ample at all times. This canoe has the most steadiness of any model we make. Stock color, Dark Green. (Another view of this model is shown on page 21.)

Length	Beam	Depth	Weight	Price	
16 ft.	34½ in.	12 in.	70 lbs.	$36.00	With keel, $37.00
17 "	35½ "	13 "	75 "	38.00	" " 39.00

Fifteen Foot—Fifty Pound "Old Town Canoe"

To meet some requirements of camping, cruising, and exploring there appears to be a limited demand for an extra-light-one-man-canoe, *i.e.*, a canoe not burdensome in portaging and at the same time capable of carrying one man and his equipment or two men with a light load. Such a canoe we have been building to order for the past eight years, so that before offering it as a stock model its worth has been tried out under most exacting conditions and diversified waterways from Newfoundland to California and from Hudson Bay territory to Florida. All materials are carefully selected for strength and correspond to the standard C. S. grade. The planking is ⅛-inch thick and the ribs are regular size, tapered at the ends and spaced 1¾ inches apart. It has a removable middle thwart for carrying. The canvas is No. 10 and finished in the same manner as the standard canoes except the final coat is of white lead paint as in the Guide's Special canoe, instead of varnish. Stock color, slate. Immediate shipment can be given without keel; with keel four days time.

Length	Width	Depth	Weight	Price	
15 ft.	34½ in.	11 in.	50 lbs.	$30.00	With keel, $31.00

"Old Town" Flat Bottom Wooden Boat—This is a well modeled, inexpensive skiff or flat-bottomed boat strongly made for rough, hard wear. It is very seaworthy. Just the right boat for fishing. Excellent for use at boys' and girls' camps and at sporting camps for general all around use. Will stand the hardest kind of service. Requires very little care. Sure to give you your money's worth if used only one season and would then probably cost you less than you could rent a boat for. With reasonable care will last for years. Stern built strong for use with outboard motor. Any number find this boat very speedy with all of the standard motors on the market. Fish well can be supplied under the middle seat for $10.00 extra. Built of native Maine white pine, painted green outside, gray inside, white gunwales, mahogany-stained seats. One pair of rowlocks with twelve-footer, and two with fifteen-footer. Plain spruce or ash oars cost $3.50 per pair for 7 ft. or $3.75 for 7½ ft. For finished oars see page 24.

Length Extreme	Width Extreme	Depth Amidships	Approx. Weight	Approx. Weight Packed	Telegraph Code Word	Price
12 ft.	41 in.	13 in.	135 lbs.	155 lbs.	Nipe	$48.00
15 "	48 "	15 "	175 "	195 "	Fiskf	50.00

In Writing Order, Give Length, Grade, Model, Color and Price. Also Extras and Equipment.

17

"Old Town" SEA MODEL

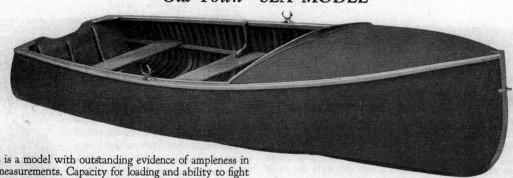

This is a model with outstanding evidence of ampleness in all its measurements. Capacity for loading and ability to fight wind and wave under adverse conditions are clearly apparent in its full bows, great depth in forward quarter, broad beam, large freeboard and flat, wide, deep stern. Bottom is modified V type. Hull is strongly braced and stiffened. Long, natural-grown knees are used. Seats are wide, three in number. Deck, constructed to bear the weight of a person, is 60″ long, of substantial wood construction, and covered with heavy canvas finished Dark Green color same as body of the boat. The bows

flare out to shoot off spray, making the boat wonderfully dry. The lines are fast, and the craft as a whole quickly draws the admiration of those who go down to the water in boats. With motors of 8 H. P. a speed of eighteen to twenty miles per hour may be attained. Smaller motors, those in the 4 H. P. class, can be counted on for about sixteen miles. Depth at stern is 16″.

Stock Color Dark Green. Other colors page 26. Price includes seats, painter ring, floor rack, outside stem, 60″ deck and one pair brass rowlocks, also crating for shipment. Bilge keels on bottom, sometimes desirable, cost $5.00. For $52.00 additional there can be furnished, as illustrated on page 23, 16″ wide seats with two back rests and three box cushions 3″ deep covered with artificial leather and filled with Kapoc. A steering wheel in bow with pulleys and eyes to carry steering rope to stern to control motor handle can be supplied for $14.00. Waterproof canvas cover with bows $20.00.

Length Extreme	Width Extreme	Depth Amidships	Approx. Weight	Approx. Weight Crated	Approx. Cubic Measurements	A. A. Grade (Page 5)	Telegraph Code Word	C. S. Grade (Page 5)	Telegraph Code Word
16 ft.	48 in.	22 in.	225 lbs.	475 lbs.	200 ft.	$175.00	Falconer	$160.00	Renocl

In Writing Order, Give Length, Grade, Model, Color and Price. Also Extras and Equipment.

"Old Town" STEP PLANE

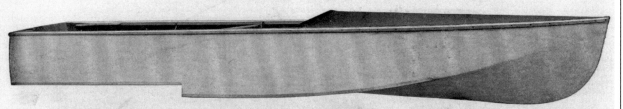

This model is primarily for racing. Its weight, about 125 lbs., is held down to the lowest point consistent with necessary strength and rigidity. Ribs, planking and all wood parts are standard dimensions, but the canvas is much lighter than on service models, while the canvas deck has only the ridge pole and end finish. Stern is braced to bottom and gunwales with long, natural-grown knees.

In lines and design this Stepper embodies the necessary and outstanding features which in similar craft during 1927 enabled speeds of twenty-five to thirty-two miles per hour with popular makes of 8 H. P. Motors. There are no seams to open up and leak under heavy pounding through rough seas—the tight canvas takes care of this. We expect to see this model win a good many races this year. It is a gem in construction and appearance.

Stock Color Dark Green. Other colors page 26. Name on two sides in 10″ to 12″ letters 40c. per letter. Price includes 72″ long deck as illustrated, painter ring, outside stem and crating for shipment. Depth at stern 15″. Brass rowlocks $3.50 per pair. Oars 7′—$4.90 per pair. Where sides and bottom meet brass angle is fastened over canvas to take the brunt of service and wear.

Length Extreme	Width Extreme	Depth Amidships	Approx. Weight	Approx. Weight Crated	Approx. Cubic Measurements	A. A. Grade (Page 5)	Telegraph Code Word	C. S. Grade (Page 5)	Telegraph Code Word
14 ft.	46 in.	17 in.	125 lbs.	375 lbs.	160 ft.	$165.00	Flashet	$150.00	Fleeter

In Writing Order, Give Length, Grade, Model, Color and Price.   Also Extras and Equipment.

21

"Old Town" BABY BUZZ

A thrill for speed exhilarates the wistful observer of this boat. It embodies lines that make water racing seem as fast as flying. Speeds of sixteen to eighteen miles are usual with 4 H. P. Motors, and 8 H. P. Motors run the boat up to twenty miles and better. However, it is more than a speed boat, although this is the characteristic that impresses you most. It has a broad, completely flat floor, except for a very slight V which gives it a steadiness unknown in other boat models, so that as an all-round general-utility boat for fishing, vacationing, family use, etc., it would be hard to find its equal when speed is an important consideration. Variable speeds of the engine permit of slow trolling travel as well as racing.

Stock Color Dark Green. Other colors page 26. Price does not include back rests, wide seats and 3″ deep Kapoc cushions with artificial leather coverings as illustrated above. These are

$52.00 additional. Price includes same type of seats as illustrated on page 22 in the Sea Model, painter ring, floor rack, outside stem, 60″ deck, brass rowlocks and crating for shipment. Steering wheel complete $14.00. Waterproof canvas cover with bows $20.00. Thirteen foot length illustrated page 24. Backs and cushions for 13′ length cost $32.00.

Length Extreme	Width Extreme	Depth Amidships	Approx. Weight	Approx. Weight Crated	Approx. Cubic Measurements	A. A. Grade (Page 5)	Telegraph Code Word	C. S. Grade (Page 5)	Telegraph Code Word
16 ft.	52 in.	22½ in.	195 lbs.	450 lbs.	180 ft.	$185.00	Honson	$170.00	Sojhn
13 "	43 "	18½ "	175 "	425 "	160 "	175.00	Ledon	160.00	Delno

In Writing Order, Give Length, Grade, Model, Color and Price.   Also Extras and Equipment.

Old Town "RACEPLANE"

Here is an all-wood step plane which has consistently shot its spray into the faces of its competitors since its debut on the 4th of last July. As an experimental model under the name "Nonsense," powered with an Evinrude Speeditwin and operated by Ellsworth Langdon of Freeport, New York, it then entered its first race, Class C, held under the auspices of Bay Park Country Club, and won. On July 21st, at Bay Head Yacht Club, it took the New York State Championship in the free-for-all. On August 25th, at Lake Ronkonkoma Boat Club races, against twenty-eight starters, it took all three heats in Class C and also won the free-for-all. Its last races of last fall were held October 21st. At Lake Ronkonkoma Boat Club it won all three heats in Class C over a four-mile course timed at over 40 miles per hour, and then took the free-for-all over a six-mile course. From a total of forty contests in which this boat competed against all popular makes of boats and motors, it took thirty-four firsts and five seconds. It never upset. Its perfect performance and easy maneuvering drew the highest praise from all who saw it. We expect astonishing things from it this season.

Built only in A.A. Grade, with planking and decking of Philippine wood (known also as mahogany) and spruce frame. Seams are battened, set in marine glue, and fastened with bolts. There are over one thousand bolts used in its construction. The hull is rigid. Brass fastenings used throughout. Cockpit comfortable, with no interfering braces. Finished in natural mahogany with spar varnish. Price includes crating. Name on two sides in 5-inch letters, 30c per letter.

The driver and his trophies

Length Extreme	Width Extreme	Depth Amidships	Height Stern	Width Stern	Approx. Weight	Approx. Weight Crated	Approx. Cubic Measurements	Price	Telegraph Code Word
11 ft., 8 in.	46 in.	17 in.	16 in.	45 in.	140 lbs.	325 lbs.	120 ft.	$225.00	Langrac

ALL-WOOD CANOE

This all-wood canoe is the finest example of the art of canoe building. No expense has been spared in perfecting design, in selecting materials, or in workmanship to assure a craft of unsurpassed quality. It is aimed to conform in measurements with racing regulations of the American Canoe Association. Rib and batten construction and finished in natural wood color with varnish. Adjustable thwarts. Price includes crating for shipment. Paddling seat can be furnished for $2.50. Double paddles priced page 34.

Length Extreme	Width Extreme	Depth Amidships	Approx. Weight	Approx. Weight Crated	Approx. Cubic Measurements	Price	Telegraph Code Word
16 ft.	30 in.	11¾ in.	64 lbs.	220 lbs.	120 ft.	$165.00	Wooderb

SQUARE-END PADDLING CANOE

There is a very definite demand for a canoe that can be used both with paddle and motor. This model has been developed and stands approved by users as an admirable all-purpose craft for the needs it is designed to fill. The general dimensions are the same as our 20 ft. "Guide's Special" but the length is cut to 18 ft. Motors not exceeding 3 H.P. are recommended. Distance between gunwales at stern is 12 inches. Stock Color Dark Green. Equipped with half ribs.

Length Extreme	Width Extreme	Depth Amidships	Approx. Weight	Approx. Weight Packed	For export (see note p. 31)		C. S. Grade (see p. 5) with keel		G. S. Grade (see p. 5) with keel	
					Approx. Weight Crated	Approx. Cubic Measurements	Open Spruce Gunwales	Telegraph Code Word	Open Spruce Gunwales	Telegraph Code Word
18 ft.	40 in.	14 in.	105 lbs.	175 lbs.	325 lbs.	160 ft.	$84.00	Moteint	$79.00	Seinotemt

In Writing Order, Give Length, Grade, Model, Color and Price. Also Extras and Equipment.

"Old Town" DINGHIES

10' 2" Dinghy (Description and Price page 18)

9 ft. Dinghy

"OLD TOWN 'SAILING' DINGHIES"

The following items added here at the factory to the 9 ft. or 11½ ft. Dinghies on page 18 produce complete sailing dinghies. To the price of the dinghy on page 18 add prices shown below. Crating $5.00.

Centerboard complete $50.00	Sail 45 ft. area, for 9 ft. Dinghy . . . $18.00	
Rudder with tiller . . 5.00	Sail 55ft. area, for 9ft. or 11½ ft. Dinghy 20.00	
Mast ring and step . . 3.00	Sail 65 ft. area, for 11½ ft. Dinghy . . 23.00	

All ready for passengers (10' 2" Dinghy)

19

"Old Town" LAPSTRAKE BOATS

Here are shown pictures of two models of Lapstrake Cedar Boats — the upper picture is a side view of the Lapstrake Square Stern Model and the lower picture of the Lapstrake Rowboat. The construction is the same in each boat but one is intended for motor use and the other for rowing. Measurements below will convey a complete understanding of the difference in beam amidships and at stern.

Superior workmanship is evident in the makeup of these boats. Planking is of selected cedar in narrow widths ⅜ in. thick. Frame is oak. Braced with natural grown knees, all fastenings brass. Complete with floor boards and rowlocks as illustrated. Finished in natural wood with varnish. For the Square Stern Lapstrake we recommend motors up to 12 H.P. All trimmings mahogany.

Length Extreme	Width Extreme	Width Stern	Depth Amidships	Approx. Weight	Approx. Weight Crated	Approx. Cubic Measurements	Price Crated	Telegraph Code Word
14 Ft. Lapstrake Square Stern	52 in.	45½ in.	18 in.	190 lbs.	350 lbs.	130 ft.	$125.00	Mestraker
14 Ft. Lapstrake Rowboat	47 "	30 "	17½ "	155 "	300 "	125 "	110.00	Straker

16

"Old Town" LAPSTRAKE SAIL BOAT

Here is a boat to delight the heart of a sailor! It's a joy to watch it perform and still more of a joy to handle the helm. Hull has lines for speed and ample beam for stability. The sail area is properly sized for speed and easy handling. For the beginner or the seasoned sailor there is the same joy of possession as the ownership of a fine automobile.

Planking is ½" white pine laid in lapstrake construction and securely fastened to ash or oak ribs and frame.

All fastenings and fittings are brass. Centerboard is wood lowered and lifted with bronze friction lever—very easy to operate. Sail is Marconi type about 68 ft. area held to mast and boom with slides. Jib has about 22 ft. area. Total area about 90 ft. Height of mast 16½ feet. Boat has 53 inch deck solidly built, also full length side decking and all covered with canvas finished in blue color. Exterior of hull finished white—interior painted light gray. Waterproof khaki canvas cover for cockpit costs $11.00.

Length Extreme	Width Extreme	Width Stern	Depth Amidships	Approx. Weight	Approx. Weight Crated	Approx. Cubic Measurements	Price	Telegraph Code Word
13½ ft.	62 in.	41 in.	24 in.	330 lbs.	500 lbs.	205 ft.	$225.00	Lapsailor

In Writing Order, Give Length, Model and Price.

26

"*Old Town*" 18-FT. LAPSTRAKE OCEAN MODEL

Here is the boat for family fishing or cruising. Big and roomy. Broad flat floor with high sides so the children can romp around with safety. Cross seat in bow and stern each easily seats five people. Flared Clipper bow for smooth dry riding. Dry well in stern with transom cut for long shaft motors for added safety. Can be furnished with standard transom height on special order at no extra cost. Plenty of room under rear seat for gas cans and battery. Front seat backrest has walkthru for ease in moving about. Large storage area under bow seat. Locker under bow deck gives added storage area. Hatch in bow deck for ease in getting to anchor. Windshield front panels can be opened. Boat comes equipped with lights bow and stern. Mooring ring in bow with chocks. Steering wheel. Automatic drain. Stern cleats.

Speeds to over 30 MPH with twin motors of 40 H.P. but performs nicely with a single motor of medium horsepower.

Clinker-built of narrow $\frac{1}{2}''$ thick cedar strakes brass screwed and bolted to ash ribs $\frac{7}{8}''$ x $1\frac{1}{4}''$. Mahogany trim and chrome brass hardware. Inside varnished. Floor boards painted with nonskid. Outside planking white with Salt Water Bronze bottom. Faded colors as shown on back cover $40.00 extra. Other accessories page 26. Shipping and storage cradle, $25.00.

Length Extreme	Extreme Width	Width Stern	Height Bow	Depth Amidships	Approx. Weight	Approx. Wt. Shipping	Price	Telegraph Code Word
18 ft. Lapstrake Ocean Model	$92\frac{1}{2}$ in.	77 in.	$46\frac{1}{2}$ in.	39 in.	1175 lbs.	1500 lbs.	$1495.00	Ostrak

In Writing Order, Give Length, Model, Color and Price. Also Extras and Equipment.

Old Town **MOLITOR MODEL CANOE**

The Molitor Model with its torpedo ends is the most distinctive canoe we offer. Long decks, outside stems and oval mahogany gunwales are included as standard equipment. With seats bolted close under the gunwales, thwarts are unnecessary. This leaves the center of the canoe open for easier loading. The Molitor is a fine craft for paddling, rowing or sailing. It will stand out in any group of canoes. Stock color Dark Green.

Length	Width	Depth	Weight	Pkd. Wgt.	Price
17 ft.	35 in.	12 in.	80 lbs.	100 lbs.	$425.00

In Writing Order, Give Length, Model, Color and Price. Also Extras and Equipment.
All prices F. O. B., Old Town, Maine.

Class "C" sail on Molitor

Bibliography

"A Novel Law Suit: The Old Town Canoe Co. Figures in Canadian Courts." *Old Town Enterprise*, 15 April 1905: np.

"A. Parker Bickmore: Originator of Bickmore's Gall Cure." *Old Town Enterprise*, 5 March 1910: 1.

"Bickmore Gall Cure Co." *Old Town Enterprise*, 28 January 1911: 1.

"B. N. Morris." *Bangor Daily Commercial*, 9 August 1912: 77.

Bureau of Industrial and Labor Statistics. Augusta, Maine. *Annual Report of the Bureau of Industrial and Labor Statistics for the State of Maine 1888–1910*. 1910.

Calloway, Colin G. *The Abenaki*. New York: Chelsea House Publishing, 1989.

_____. *Dawnland Encounters*. Hanover: University Press of New England, 1991.

"Canoe Company Would Do Well in Old Town." *Old Town Enterprise*, December 1894.

Cardin, Bob. *Old Veazie Railroad, 1836, One of America's Very First Railroads*. Bangor, ME: Galen Cole Family Foundation, 1992.

Cheever, David. "Old Town Canoe Has New Kayaks." *Bangor Daily News*, 21 March 1979: np.

"The City, Old and New Industries." *Old Town Enterprise*, 27 April 1911: 1.

Devitt, Mary Josephine Orr. *Selected Aspects of the Development of Old Town, Maine*. Thesis. Orono, ME: University of Maine Press, 1949.

Eckstorm, Fannie Hardy. *Indian Place-Names of the Penobscot Valley and the Maine Coast*. Orono, ME: University of Maine Press, 1978.

_____. *The Penobscot Man*. Somersworth, NH: New Hampshire Publishing Company, 1972.

Gray, S. B. "How We Built New Markets for an Old Product." *System: the Magazine of Business*, January 1927: 56–58.

Hempstead, Alfred. *The Penobscot Boom and the Development of the West Branch*. Orono, ME: University of Maine Press, 1931.

History of Penobscot County Maine, with Illustrations and Biographical Sketches. Cleveland: Williams, Chase & Co., 1882.

Johansen, Jon. "From Gray's Hardware, the Old Town Was Born." *Bangor Daily News*, 29 February 1984: np.

_____. "OT's 'Discovery' Leads Canoe Industry." *Bangor Daily News*, 7 March 1986: np.

Johnson, Samuel C. *The Essence of A Family Enterprise*. Indianapolis, IN: The Curtis Publishing Company, Inc., 1988.

Joy, Arthur F. "A Canoe, the Charles River, and You." *Yankee Magazine*, October 1957.

Leading Business Men of Bangor, Rockland, and Vicinity. Boston: Mercantile Publishing Company, 1888.

"Lumber Camps of Many Years Ago." *The Penobscot Times*, 27 March 1941: np.

Manry, Robert. *Tinkerbelle*. New York: Harper Row Publishers, 1966.

McCormack, Tim and Alan Hirsch. "A 'Discovery' Revitalized Old Town Canoe Company." News Release, Phillips 66 Company, 10 August 1987: np.

"Mill Burned at Old Town." *Old Town Enterprise*, 18 May 1911: 2.

Murray, William. "Reminiscence of My Literary and Outdoor Life." *The Independent* [New York], 1904: 278.

Murray, William H. H. *Adventures in the Wilderness*. New York: Syracuse University Press, 1989.

"My Old Town Canoe." *Old Town Enterprise*, 25 December 1921: 6.

Norton, David, Esq. *Sketches of the Town of Old Town, Penobscot County Maine, From its Earliest Settlement to 1879*. Bangor, ME: S. G. Robinson, 1881.

"Old Town Canoe Co." *Old Town Enterprise*, 25 February 1911: np.

"Old Town Canoe Company: Old Factory, New Ideas." *Bangor Daily News*, 30 January 1971, sec. B: 16.

"Old Town Canoe Company—One of the Most Phenomenally Successful Industrial Enterprises in Eastern Maine." *Old Town Enterprise*, 3 March 1905: 1.

"Old Town Canoe Strike At End." *Bangor Daily News*, 22 May 1954: np.

"Old Town Firm to Build Canoes For Disneyland." *Bangor Daily News*, 24 April 1956: 6.

Old Town, Maine, The First 125 Years, 1840–1965. Maine: Penobscot Times, 1965.

"Old Town Man Sues Fredericton Canoe Maker." *Bangor Daily News*, 12 April 1915: 3.

"Old Town, the Canoe Center of the World." *Old Town Enterprise*, 19 March 1910: np.

"Penobscot Indians." *Penobscot Times*, 21 August 1941: 4.

Perry, Ralph Frederick. *Canoeing the Charles—Images and Field Notes from 1902–12*. Hollis, NH: Hollis Publishing Company, 1996.

"Pickets Are Stationed At Canoe Plant." *Penobscot Times*, 25 March 1954: np.

Pollock, Robert. *Out To Norumbega*. Unpublished manuscript, 1996.

Punnett, Dick. *Racing on the Rim*. Ormond Beach, FL: Tomoka Press, 1997.

"Rafting on the Penobscot Shows Sharp Decline in 22 Years." *Old Town Enterprise*, 25 November 1922: 1.

Richardson, James D. *A Compilation of the Messages and Papers of the Presidents 1789-1897*. Washington: Government Printing Office, 1896.

Stanton, G. Smith. *Where the Sportsman Loves to Linger*. New York: J. S. Ogilvie Publishing Company, 1905.

Stelmok, Jerry and Rollin Thurlow. *The Wood & Canvas Canoe: A Complete Guide to its History, Construction, Restoration, and Maintenance*. Gardiner, ME: Harpswell Press, 1987.

"Strikers Picket Old Town Canoe Plant." *Bangor Daily*

News, 23 March 1954: 5.

Thoreau, Henry David. *Canoeing in The Wilderness*. Boston: Houghton Mifflin Company, 1916.

_____. *The Maine Woods*. New York: Thomas Y. Crowell Company, 1961.

"Today is 121st Anniversary of Big Real Estate Deal." *Penobscot Times*, 29 June 1939: 1.

Vetromile, Reverend Eugene. "Pamola." *White Pine and Blue Water: A State of Maine Reader*. Edited by Henry Beston. Camden, ME: Down East Books, 1950.

Chapter Opening Credits

FRONT COVER: *Legacy*, a 1953 OLD TOWN War Canoe owned by Sue and Vincent Audette. Photograph by Benjamin Mendlowitz copyright © 1998.

BACK COVER: OLD TOWN CANOE COMPANY catalog covers from 1915 (left) and 1917.

TITLE PAGES: Photograph of *Legacy*'s interior by Benjamin Mendlowitz, copyright 1998; 1928 and 1932 OLD TOWN CANOE COMPANY CATALOG covers.

CHAPTER 1 OLD TOWN CANOE COMPANY catalog, 1915

CHAPTER 2 Detail of Carleton shop in late 1880s, p. 13.

CHAPTER 3 Detail from INDIAN OLD TOWN CANOE COMPANY catalog, circa 1901.

CHAPTER 4 Detail of Indian guide, p. 25.

CHAPTER 5 Detail of Norumbega Park ad, p. 32.

CHAPTER 6 Detail of OTC staff photo, p. 39.

CHAPTER 7 OLD TOWN CANOE COMPANY catalog, 1910.

CHAPTER 8 Detail from PECACO Canoe Company catalog, circa 1920.

CHAPTER 9 OLD TOWN CANOE COMPANY, 1915.

CHAPTER 10 Detail of Pearl Cunningham, p. 76.

CHAPTER 11 OLD TOWN CANOE COMPANY catalog, 1946.

CHAPTER 12 FG model canoe, from 1968 OLD TOWN CANOE COMPANY catalog.

CHAPTER 13 OLD TOWN CANOE COMPANY catalog, 1968.

CHAPTER 14 OLD TOWN CANOE COMPANY catalog, 1972

CHAPTER 15 OLD TOWN CANOE COMPANY catalog, 1998. The inset on page 127 is from the 1930 OLD TOWN CANOE COMPANY catalog cover.

About the Authors Susan Audette began canoeing as an adult and quickly became an enthusiast. She was a member of the first United States Canoe and Kayak Marathon Team to attend the World Championships (1988), and although her love of the sport includes sleek racing canoes and kayaks, wood-and-canvas canoes remain a favorite. With her husband Vincent, she operates a retail canoe and kayak shop called the Water Works Canoe Company in North Windham, Connecticut. Sue is an active member of the Wooden Canoe Heritage Association, as is her co-author, David Baker, who has been canoeing since childhood. Dave collects and restores canoes, has served as historian and librarian of the WCHA and is now its current president, and consults, teaches, and lectures for museums and organizations.